PARADISE LOT

PARADISE LOT

TWO PLANT GEEKS, ONE-TENTH OF AN ACRE

=== *and* ===

THE MAKING OF AN EDIBLE
GARDEN OASIS IN THE CITY

ERIC TOENSMEIER

WITH CONTRIBUTIONS FROM JONATHAN BATES

Chelsea Green Publishing
White River Junction, Vermont

Developmental Editor: Brianne Goodspeed
Project Manager: Hillary Gregory
Copy Editor: Alice Colwell
Proofreader: Helen Walden
Indexer: Lee Lawton
Designer: Melissa Jacobson

Printed in the United States of America.
First printing January, 2013.
10 9 8 7 6 5 4 3 2 1   13 14 15 16

Our Commitment to Green Publishing

Chelsea Green sees publishing as a tool for cultural change and ecological stewardship. We
strive to align our book manufacturing practices with our editorial mission and to reduce the
impact of our business enterprise in the environment. We print our books and catalogs on
chlorine-free recycled paper, using vegetable-based inks whenever possible. This book may
cost slightly more because it was printed on paper that contains recycled fiber, and we hope
you'll agree that it's worth it. Chelsea Green is a member of the Green Press Initiative (www.
greenpressinitiative.org), a nonprofit coalition of publishers, manufacturers, and authors
working to protect the world's endangered forests and conserve natural resources. *Paradise Lot*
was printed on FSC®-certified paper supplied by Thomson-Shore that contains at least 30%
postconsumer recycled fiber.

Library of Congress Cataloging-in-Publication Data
Toensmeier, Eric.
  Paradise lot : two plant geeks, one-tenth of an acre, and the making of an edible garden oasis
in the city / Eric Toensmeier ; with contributions from Jonathan Bates.
      p. cm.
  Includes bibliographical references and index.
  ISBN 978-1-60358-399-2 (pbk.) — ISBN 978-1-60358-400-5 (ebook)
1. Urban gardening—Anecdotes. I. Bates, Jonathan, 1974– II. Title. III. Title: Two plant
geeks, one-tenth of an acre, and the making of an edible garden oasis in the city.

  SB453.T59 2013
  635.09173'2—dc23
                                    2012043044

Chelsea Green Publishing
85 North Main Street, Suite 120
White River Junction, VT 05001
(802) 295-6300
www.chelseagreen.com

CONTENTS

Part Four: Reap
(2009–2012)

PART ONE

SLEEP

2000–2004

1

GERMINATION

Few gardeners would have looked out at the small, flat expanse of compacted fill and thought, "It's perfect!" The front yard was a short, steep slope of asphalt with a tiny strip of sterile gravel and subsoil. Two shady side alleys led to a backyard that looked like a moonscape, sparsely populated with tufts of crabgrass. Two large Norway maples, reviled as weeds that poison anything growing beneath them, hung over the garden from the north side. And the house itself was soulless, all straight lines, devoid of personality. It was exactly what we were hoping for.

As we walked this damaged piece of ground, my friend Jonathan and I imagined a lush paradise of fruits and berries, interspersed with ponds, greenhouses, and bountiful beds of annual crops. This would be a perfect place to test our ideas. Could we bring this land back to fertility and ecosystem health? Could we garden every square inch, creating a diverse and beautiful edible landscape? Would these and other permaculture strategies, proven in Australia and the tropics, work in Massachusetts, with our short summers and frigid winters, enabling us to harmonize our goals with the needs of this misused and neglected piece of land?

To properly tell the story of our experiment—why Jonathan and I thought a barren urban lot held such potential—I first have to back up and tell the story of how we got there.

After I graduated from high school, I didn't feel ready for college. Instead, in 1990, at the age of nineteen, I went to intern at the Schuylkill Center for Environmental Education (SCEE), an urban nature center close to my home in Philadelphia. As a ten-year-old, I had participated in SCEE's weeklong Sunship Earth program. On the first day I chose my "magic spot," a special corner of the meadow where I went each day to spend fifteen minutes quietly observing and recording those observations, writing thoughts, and drawing pictures of the flowers, grasses, and insects around me. This simple exercise I did as a child sparked my curiosity and wonder for the natural world for years to come. (I still use it with students, young and old, in the permaculture classes that I teach.)

When I returned to SCEE as an intern, "magic spots" had become a Lyme disease hazard, but innovative environmental education continued. I worked for a woman named Lori Colomeda, who had run the Sunship Earth program I had attended ten years earlier. Though the other interns all had bachelor's or master's degrees, I think Lori took a chance on me because I had come through Sunship Earth. And so she changed my life twice. Lori took our intern team to a one-day environmental education conference nearby, where I took part in a workshop on permaculture being offered by Bob McKosky, an early adopter of permaculture in the United States. I don't remember what he said in that sterile classroom, but I do remember that I went right back to the SCEE library after the conference and read straight through Bill Mollison and David Holmgren's *Permaculture One*, a book that set the course for the rest of my life.

Permaculture (short for "permanent agriculture" or "permanent culture") is a movement that began in Australia in the 1970s. It brings together traditional indigenous land management practices, ecological design, and sustainable practices to create landscapes that are more than the sum of their parts. Permaculture is not so much about having a greenhouse, chickens, and an annual vegetable garden as it is about how those elements are tied together to create functional interconnections that work like a natural ecosystem. Low maintenance is a holy grail in permaculture—a food forest with a hammock hidden beneath fruit trees, where, as permaculture codeveloper Bill Mollison famously quipped, "the designer turns into the recliner."

Permaculture teacher and designer Rafter Ferguson came up with one of my favorite definitions of permaculture: "meeting human needs while improving ecosystem health."

This idea of not passively observing ecosystems but actively engaging with them, designing and coevolving with them, had a huge impact on me when I was a teenager at SCEE and has stayed with me to this day. About as far as I had got in environmentalism until that point was to try to stop doing bad things to the environment—to minimize my negative impact. For the first time, I was exposed to the idea that people could have positive impact, that by interacting with the land we could benefit both ourselves and the environment.

I was also captivated by the notion of food-producing ecosystems, with multiple layers like a forest but growing food at the same time. *Permaculture One* was filled with diagrams of food forests, nut pines and nitrogen-fixing tagasaste trees towering over feijoa (pineapple guava) bushes, tree tomatoes, and hazelnut shrubs; passionfruit vines climbing everywhere; and an understory of comfrey and sweet potatoes and ginger. Sounded pretty fantastic to me!

Despite my newfound passion, I had a dilemma: permaculture wasn't taught at most colleges at that time. I was attending Hampshire College in Amherst, Massachusetts, with the hope that their independent learning program would allow me to study what I wanted. My classes didn't teach me what I was keen to learn, so I spent hours in the library reading the books that Mollison and Holmgren had read when they were writing *Permaculture One*. I immersed myself in James Lovelock's *Gaia: A New Look at Life on Earth,* J. Russell Smith's 1927 masterpiece *Tree Crops: A Permanent Agriculture,* Victor Papanek's *Design for the Real World,* Ervin Laszlo's *Systems View of the World*, and everything I could find by the New Alchemy Institute. But even though I was learning so much, I failed to work with my adviser to get credit for this independent research and fell between the cracks academically.

By the middle of my second year of college, I knew that what I really needed I could get only from a working permaculture farm.

Around that time I went to an introductory permaculture workshop taught by Dave Jacke. He was lean and intense, a passion for permaculture burning in his eyes that I understood all too well. Little did I know how closely Dave and I would work in the years to come to take permaculture in the United States to another level. Dave was selling copies of Mollison's 600-page *Permaculture: A Designers' Manual*. I persuaded a friend to loan me the money to buy it, then read it from cover to cover. Around the time I reached page 200, I was on a road trip through a beautiful valley in Vermont and envisioned every hillside terraced and covered in tree crops. I've never been able to look at a landscape the same way again.

During my last semester at Hampshire, a graduating senior came to speak to my urban ecological design class about Nuestras Raíces (Our Roots), a project that had begun life as a community garden in the low-income Puerto Rican neighborhoods of Holyoke. Almost everyone in the older generations in that community had grown up in the mountains of Puerto Rico. They were raised on farms and had worked as farm laborers in sugar cane back home and vegetable and tobacco farms up and down the East Coast of the United States. They were people with deep agricultural roots living in a city of brick and cement. I was deeply inspired but could not have conceived that I would spend much of my adult life involved in Nuestras Raíces in one way or another.

Since it was clear that college was a dead end for me, that I needed instead an experiential education, I mailed a letter to every permaculture site listed in the back of Mollison's *Designers' Manual* and asked for an internship. In May 1992 I set off to spend six months with Jerome Osentowski at the Central Rocky Mountain Permaculture Institute (CRMPI) in the high desert of Colorado.

At CRMPI I saw all of the principles of permaculture that I'd read about. Jerome's greenhouse was a marvel. Excavated into the south slope of a mountain, it enabled year-round production despite the harsh winters, which reached −20°F (and chilly summer nights as well). Compost piles along the north wall provided insulation and heat generated by the decomposition of organic matter. At night chickens were closed up in a chicken house that Jerome built on the west side of the greenhouse; he had calculated how many BTUs each chicken gave off at night and could tell you the insulation and heating value of his poultry house. Water tanks along the northern wall inside the greenhouse absorbed heat during the day and radiated it back at night. Low-energy electrical fans moved hot air from the ceiling of the greenhouse and pumped it through perforated pipes under the garden beds, radiating heat into the soil. At night the fans reversed and pumped out the heat to warm the crops. A sauna built on the north side of the greenhouse warmed both people and plants on chilly nights. Any of these fairly passive strategies alone would not have been enough, but together they combined to do the job.

Jerome's chicken yard was also a wonder. There was a gate at the top of the steep slope through which we tossed weeds, food scraps, and straw. The chickens ate and shredded and scratched at this material, slowly moving it down the hill and maneuvering it all the way. By the time it reached the bottom of the hill, it was magnificent compost. Over the course of the summer, it accumulated to several feet deep, which we dug out and put in a pile to cook for a few weeks before spreading it on the vegetable beds. Jerome's chickens were happy and healthy and had plenty of sunlight and fresh greens and insects to eat, but they were also working for the farm. He had designed his system so that the chickens—just by expressing their innate behaviors—would make less work for him while making nutritious eggs and meat. I was entranced.

During the summer I spent at CRMPI, our intern crew installed stone terraces and planted some of the first trees that today make up one of the oldest forest gardens in the United States. It was at Jerome's that I tasted my first heirloom tomatoes fresh off the vine, ate pinyon pine nuts collected in the forest, and fell in love with salads of the highest-quality greens. I learned that a truly fresh salad needs no dressing and developed a lifelong snooty attitude toward salad greens.

When I returned to Massachusetts in October 1992, I resolved to go after permaculture with everything I had. I left Hampshire and enrolled in a bachelor's program at the Institute for Social Ecology in Plainfield, Vermont, that allowed me to deeply pursue permaculture as an independent study. I worked on restoring part of a neglected old apple orchard in Easthampton, Massachusetts, using the edible forest gardens model. I was also interning at Tripple Brook Farm, a nearby nursery with a world-class collection of useful plants. As the master (or perhaps servant) of a collection of fifteen hundred species of useful plants, the owner, Steve Breyer, became my botanical mentor. I planted an understory for the apples that included perennial vegetables like Jerusalem artichoke, rhubarb, and fuki, along with bee balm and strawberries and other interesting crops. Even though I wouldn't plant those today, it was an early step on a long journey. I learned that without intensive spraying (even of organic sprays), an orchard of one hundred apple trees won't make a single apple worth eating. It's hardly a low-maintenance crop! Meanwhile, most of the pear trees were bearing beautifully—a valuable lesson.

In the process of working on the orchard and writing my thesis, I discovered that most of the resources on useful perennial plants for permaculture applied to tropical or subtropical climates—not much use in Massachusetts. From the beginning of my interest in plants for permaculture and edible landscaping, I identified perennial vegetables as a gap in available information. Nobody seemed to know which perennial leaves, roots, and shoots could grow under or between fruit trees, nut trees, and berry bushes. People are sometimes shocked to learn that we have a fine palette of long-lived and low-maintenance vegetable crops for cold climates, many of which are also fine ornamentals. Little did I know as I worked on my first list of useful cold-hardy perennial species for the appendix of my thesis that I would spend much of the next two decades filling in the gaps of knowledge and developing an insatiable appetite for exploring the useful plants of the world.

For several years after I finished my coursework at the Institute for Social Ecology, I pieced together a living with educational if not lucrative part-time jobs, like collecting seeds of native wetland plants and tending the basil at a large aquaponics operation (integrated fish farming and hydroponics in a recirculating system). My adviser from the Institute for Social Ecology, the great Grace Gershuny, recommended me for a position as the custodian of their sustainable agriculture library at the New England Small Farm Institute (NESFI) in Belchertown, Massachusetts. The first day I showed up to work, NESFI's director, Judy Gillan, introduced me to their collection. The books were loaded into the back of a farm trailer, covered by a tarp with a bit of snow on top of it, parked in the barn they were turning into their office. The time I spent organizing this library was an unparalleled learning opportunity. Every month I filed the hundred agriculture magazines NESFI received and read the classics that underlie permaculture and the modern organic movement. I made my way through Sir Albert Howard's *Agricultural Testament,* Lady Eve Balfour's *The Living Soil,* P. A. Yeomans's *Water for Every Farm,* Edward Faulkner's *Ploughman's Folly,* volumes by Wes Jackson and Wendell Berry, and F. H. King's *Farmers of Forty Centuries, or Permanent Agriculture in China, Korea, and Japan.* NESFI's library is a treasure, and I encourage anyone passing through Massachusetts to visit it. It's worth traveling to New England just to spend a few days there.

During this period I also began to do a bit of permaculture teaching and design work, which brought me back in contact with Dave Jacke. We had both been asked to bid on a teaching gig. Dave was a hundred times more qualified than I, but rather than take the job for himself, Dave called me up and asked if I would like to do it with him. We spent an hour on the phone, quickly recognizing our shared passion for permaculture. We organized our first weekend edible forest gardens workshop at the New England Small Farm Institute in 1997.

During that workshop, Dave and I were approached by Ben Watson, an editor at Chelsea Green, about writing a book. We set out to write a short introduction to forest gardening for adventurous gardeners. Eight years and over a thousand pages later, *Edible Forest Gardens* was released in two volumes.

My primary role as junior coauthor was to develop tables of useful cold-hardy perennial species, a job I delved into to the point that I eventually developed a list of more than six hundred useful perennial species for the eastern forest region. Dave Jacke brought something totally different to the project. Dave is a design genius, a man with incredible knowledge about a broad range of topics from advanced ecology to grading parking lots. Working on the book with him allowed me to investigate useful species systematically while learning about ecology and design process. As the book developed, we realized we were mapping out a grand hypothesis while issuing a challenge to the permaculture community: let's see if forest garden systems are viable in the eastern United States.

Of course that challenge was one I really wanted to take on myself. As a writer during my late 20s and early 30s, I had little money and no land. I was working with Dave on *Edible Forest Gardens* writing about luscious landscapes and edible fruits while living in small apartments with no gardens of my own. I had no Asian pears or persimmons or hardy kiwis. And I had no one to grow and enjoy them with. It was embarrassing to be teaching with little hands-on experience, but more importantly I simply longed to have those plants part of my life. For years, I went to bed in those apartments rereading chapters of Lee Reich's *Uncommon Fruits Worthy of Attention,* pining for the day when I would taste my own medlars, kiwis, and pawpaws.

I had another major challenge to deal with. In October 1994 I had sustained a fairly serious head injury, followed over the next decade by lesser additional head injuries. The pain made it difficult to concentrate. As I described it to my friends, it was as though my brain was a bloated groundhog crashed out on the bottom of my brainpan, while my brainstem felt like a wilted zucchini. I had a migraine for three straight years and then another eleven years of constant headache.

The bright spot in my day was studying plants, which seemed to be the only thing my brain was capable of doing. In science fiction, people with brain trauma use visualizations to remap the layout of their brains. As unbelievable as it sounds, that's what I did with botany.

I learned not only the Latin names of useful species but the families to which they belonged. I further memorized the orders into which families are clustered, the superorders to which orders belong, and the even larger divisions between monocots and dicots (the flowering plants) and the pre-angiosperms like ferns, ginkgo, and pines. My constant guide in this effort was Alan Kapular and Olafur Brentmar's *Coevolutionary Structure for the Plant Kingdom.* With its help I was able to categorize and file away thousands of plant names in my head.

I also undertook a project that reflects how pathetic my social life was at the time. I tracked down a copy of Steven Facciola's magnificent *Cornucopia: A Sourcebook of Edible Plants,* a remarkable compendium of three thousand species of edible plants fully cross-indexed with nursery and seed company sources. I wanted to systematically comb through this to increase my knowledge of hardy perennial food plants. I spent weekend nights at the Smith College Science Library with my copy of *Cornucopia* and the library's reference materials (primarily Liberty Hyde Bailey's *Hortus Third*) to determine which species profiled in *Cornucopia* were cold-hardy and perennial.

To organize my thoughts, I went through *Cornucopia* superorder by superorder and kept my notes accordingly. I learned a number of interesting things as a result. For example, the Magnolia superorder (now reclassed as the Magnoliales but with largely the same cast of characters) contains many trees and shrubs with aromatic leaves used as bay leaves (including bay laurel, the "official" bay leaf), as well as trees and shrubs that have exclusive relationships with certain butterflies whose larvae can eat only leaves from that particular species. These woody plants are from different families, and without an understanding of their higher-level taxonomic relationships, I would never have made that connection. Studying plants at the superorder level was part of my recovery from my brain injuries and introduced a lifelong love of classification. It also formed the foundation of the plant matrices I developed for *Edible Forest Gardens,* which were to fill many years of my life to come.

2

STARTING A SEED COMPANY
AND GAINING A GARDENING COMRADE

My love of perennial food plants and edible ecosystems spawned a number of business ideas throughout the late 1990s. I had a vision of operating a multilayered farm, with Asian pears and chestnuts growing above berry bushes and an understory of medicinal herbs and perennial vegetables, selling my harvest at upscale farmers' markets and to fancy restaurants. But it's hard enough to market produce that people are familiar with, let alone strange new crops, so I thought that selling the seeds of perennial vegetable plants might be a more successful venture.

Though I had no savings and little income, I wanted to find a place to live with enough land that I could grow the plants I was researching for *Edible Forest Gardens* and harvest the seeds from some of those plants to sell. In September 2000 I moved out of one of my crummy apartments and into a funky old farmhouse in the country just outside of Holyoke, in Southampton, Massachusetts. The historic farmhouse—which in the 1970s had been extensively redecorated with fake brick paneling and shag carpets—was situated in a rolling landscape of green hills that had once been productive farmland and was now rapidly being developed.

It wasn't long after moving to the farmhouse that I knew I needed help with the seed company. Jonathan Bates was one of my students during a three-week summer intensive course in ecological design that I taught at the Institute for Social Ecology. Jonathan was in a one-year master's degree program to research the potential for municipal-scale composting toilet systems. For his thesis, he created a "living machine" to treat greywater from a building on the campus, a complex of pipes and tanks with a riot of plant growth exploding outward, processing and filtering the bacteria and excess nutrients from the water. My kind of guy.

Older than the other students, Jonathan was turning his life upside down by leaving a career in marine aquaculture to do more practical ecological projects. He seemed serious and introspective, and by the time I was searching for someone to help me get this garden going, Jonathan was a freshly minted graduate looking for a project to plug into. He moved into the farmhouse as my "intern" for the 2000 season of the seed company. We soon discovered that he was going to be a full partner in the larger project of field-testing edible forest gardens and producing seeds for our catalog.

Our landlord, Charlie, had grown up across the street, and his family had been farming in the neighborhood for generations. Charlie plowed up a third of an acre of garden space for us before the ground froze, and that winter Jonathan and I began planning an ambitious garden.

We were renting and didn't know how many years we would be at the farmhouse in Southampton; though it was a bit heartbreaking, I knew that planting trees and even shrubs at this location would mostly be a waste of money, so we decided to focus on herbaceous perennials and resolved to take full advantage within the constraints the situation offered: good soils, availability of manure and mulch, and a garden we could see out of the big bay windows in the living room. I found, as I often have, that constraints provide not only a good challenge but a solid hook for a design.

We wanted to grow perennial vegetables and other edible perennials and see how they tasted and yielded. And we wanted to trial plants for beneficial insects and see if pest-control strategies I researched would work for us. Finally, we wanted to grow some annuals and just be able to eat our own fresh, organic food. It was time for sowing the seeds of plants I hoped I would one day take with me to my "real" forest garden when I owned land of my own and to try out for myself the things I had learned from my mentors.

It was fantastic to have Jonathan to work with. We were both wildly enthusiastic about plants, so there was always plenty of energy to carry the project forward. Jonathan also has a much stronger back than I do and was able to do a lot of the physical labor that was beyond my capacity. Because it was on White Loaf Ridge, we called our operation Wonder Bread Organic Farm.

Though Jonathan and I would grow many annuals, such as kale, tomatoes, and sweet corn, perennials such as sea kale and good King Henry were to become the backbone of our style of gardening. For me, *Permaculture One* had driven home the essential role of perennial food plants in permaculture. Perennials build soil and control erosion, improve rainfall capture, and sequester carbon. Once established, they return year after year, providing yields with little work. It seemed to me that with the right balance of low-maintenance perennials in the landscape, we might need to do very little planting and maintenance and would instead be able to focus our time and energy on harvesting. Having worked on annual vegetable operations and experienced the hard labor of planting and caring for annuals, I considered low-maintenance edible perennial vegetables an appealing alternative.

Even more than low maintenance, however, we loved perennials that were low maintenance *and* performed multiple functions. For example, we were interested in goumi, a bush that not only fixes nitrogen but produces small, tasty cherries around the same time that strawberries come into season. Sweet cicely is an ornamental that attracts beneficial insects for pest control while producing large seeds that taste like licorice. Strawberries are a great example, too, serving as a groundcover and making excellent edible fruit.

Since we were renting, it wasn't realistic to have a full forest garden with tree, shrub, and herbaceous layers. But we could experiment with high, medium, and low herbs and running groundcovers below those. As this was the aspect of forest gardening that other people had least explored for cold climates and it was at the core of my research, it seemed like a satisfying enough challenge for a few years.

That winter Jonathan and I ordered seeds of perennial vegetables from unusual companies from around the world, including the French B&T World Seeds, which offers more than thirty-two thousand species and varieties. The seeds and shipping were not cheap—sometimes coming to twenty dollars for just ten seeds—but we only had to order and plant these crops once: that's another great thing about perennials. Many of those seeds turned into plants that are with us more than a decade later, still bearing crops.

Our house had an east-facing, glassed-in porch, which we converted into a makeshift greenhouse in the early spring of 2001. We insulated it, built some shelves, set up a watering system (a simple hose), and rigged up a space heater for cold nights. It worked well enough, and we were excited about it, but even then we knew that it was no substitute for a real greenhouse. For the time being, it would have to do.

Our wild and rare perennial seeds arrived. What were we most excited about? Skirret is like a perennial parsnip that also attracts beneficial insects; groundnut, a native root crop that fixes nitrogen; good King Henry, a perennial spinach for partial shade; plus perennial scallions like Welsh and walking onions. We had hopes for some perennial brassicas like tree collards and nine star broccoli, though sadly they would consistently perish in our winters. We even tried a number of perennial grains, which also failed to overwinter or perennialize. Among these were perennial forms of corn, sorghum, rye, wheat, and something called agrotriticum, a cross between the superweed quackgrass and annual wheat. Again, no success, but it was fun trying.

Unlike annual vegetables, the somewhat obscure perennials we were growing are not fully domesticated. In many cases you can't just plant a seed and expect it to germinate in a few days. Some take more than a month to sprout. Some require stratification, a cold, wet period in the refrigerator that imitates the wet winters of another climate. For others, you must imitate the digestive process of their dispersal agent in their homeland, perhaps a bird with strong digestive juices. These seeds must be scarified—nicked with a knife or file, immersed in acid, or treated with boiling water to break their seed coat and stimulate them to come out of dormancy. I reminded myself that this was a process we would need to do only once.

Jonathan and I obviously weren't interested in growing vast fields of any single species. We wanted to grow multiple species together in polycultures, or combinations of species that minimize competition and bring out beneficial relationships. Polycultures have been shown by scientists to keep many pest populations from exploding. The best-known and best-researched polyculture in cold climates is the traditional three sisters of corn, beans, and squash, grown from Mexico to southern Canada for over a thousand years. Using the right techniques and the right varieties, this polyculture can yield more than monocultures of any of its component species. Though the plants compete for water and light, they can also help each other by fixing nitrogen, controlling pests, and suppressing weeds. In the tropics, commercial-scale polycultures are common: coffee is grown under nitrogen-fixing ice cream bean trees with edible pods; black pepper vines climb edible palm trees. But for all of the time I had spent researching, I found few examples of polycultures using perennial plants in cold climates. There was work being done on raising mushrooms, ginseng, and other woodland medicinal plants under sugar maples. While this sounded interesting, Jonathan and I daydreamed about more complex and nutritious polycultures.

Jonathan and I created a polyculture composed of three perennial root crops. It began in part as a perennial version of the three-sisters polyculture. Tall Jerusalem artichokes take the place of corn and attract beneficial insects; climbing, nitrogen-fixing groundnuts fill the role of beans. We didn't have a perennial squash, so we used Chinese artichoke, a spreading mint relative that would cover the ground quickly like squash. Though not perfect, this was one of the most successful combinations of species that Jonathan and I developed while living at Wonder Bread.

Jonathan and I were not excited about spraying our garden with pesticides, organic or otherwise. But we also didn't want to spend a lot of time squishing Colorado potato beetle larvae by hand. So we tried to think about how to prevent pest problems before they happened. We had a leg up because we had chosen crops that have few pests in the first place. We also provided food and habitat for beneficial insects and other animals that prey on pests; ponds and rotten logs, for example, help make homes for frogs and toads.

But of particular interest to Jonathan and me was planting a season-long flowering calendar of plants from the aster and umbel families to keep predatory parasitoid miniwasps fueled as they hunted for caterpillars, aphids, and other pests to eat or lay their eggs in. At Wonder Bread we had success with this technique: by late summer all of our tomato hornworms had little white cocoons on their backs, a sign that they were being slowly eaten from the inside by tiny larvae hatched from eggs a wasp had laid inside the living caterpillar.

Our Wonder Bread garden was mulched and mostly no-till. We used only organic fertilizers and pesticides. All those groundcovers and mulches help protect the living soil on which we depend.

Throughout our garden we planted nitrogen-fixing plants and other fertility-improving species, like comfrey and lupines, to provide free fertilizer for their neighbors as their leaves and root hairs slowly decomposed. Nitrogen-fixing plants have a symbiotic relationship with bacteria (and sometimes actinomycetes, yeastlike microorganisms) in their roots. The plants trade carbohydrates from photosynthesis for nitrogen, which the bacteria (but not the plants) can pull from the atmosphere. Nitrogen-fixing plants essentially make their own fertilizer, which becomes available to their neighbors over time. It's free fertilizer for anyone patient enough to wait a few years.

I loved those three years at Wonder Bread, where I could see the stars at night and watch the wood ducks perched in the maple tree above the driveway. Jonathan and I could hike out the back door and walk up onto White Loaf Ridge. Up in those woods of oak and hemlock, we learned to identify wild edible mushrooms and found several great scores of chicken of the woods. Imagine twenty pounds of tender, fluorescent orange mushroom that tastes like chicken breast!

From the top of the ridge, Jonathan and I would sometimes hike down into the valley to see Steve Breyer at Tripple Brook Farm. Our favorite time to visit was fall, when we could sample hardy kiwifruit, chestnuts, and American persimmons. One spring, after an unusually hard winter, many of Steve's thirty species of hardy bamboo were frozen to the ground, leaving dead but useful culms. We collected a huge bunch of bamboo poles and tied them up, hauling the fifteen-foot-long bundles home on our backs. I felt as if I had gone another level deeper—walking instead of driving, using local materials, getting rooted in a place enough to know where the bamboo grew and what the winter had done to it this year. Carrying bundles of bamboo poles on my back, I also felt as if we'd been transported to medieval China.

I loved the greenhouses at Tripple Brook Farm. I was getting inspired by Eliot Coleman's *Four-Season Harvest*, one of the many books on farming and gardening that changed my life. Coleman's system allows production of salad vegetables all winter long without using any heat. In true permaculture fashion, Coleman relies on multiple strategies, each of which requires minimal energy input. The first layer is the greenhouse. An ordinary hoop-frame house works just fine. Inside the greenhouse, over the beds, you build a mini greenhouse. The final and essential trick is that you can't grow just any old vegetable: you must choose species that are adapted to short days and cold—no tomatoes or okra in February.

Steve put his engineering degree to good use by developing some clever and innovative designs for low-tech greenhouses. One is built onto the side of his house and stays frost-free year-round due to insulation and heat provided by the house. He also has several freestanding greenhouses that are well insulated though unheated. Featuring double layers of plastic and insulated north sides, Steve's greenhouses stay warm enough to produce fresh figs in January.

Unfortunately, practicing any long-lived techniques on rented land has its challenges. By this point in my life, I'd planted three or four edible forest gardens in different locations. Every time I did it, I knew that in a few years I would be gone and no one would be there to care for the young edible ecosystem I had planted. It was fun the first few times because I learned a lot working hands-on with the plants, but it lost its charm as I moved from apartment to apartment on my writer's wages. This was the reason I moved to Wonder Bread and recruited Jonathan to join me. I wanted to grow every useful herbaceous perennial I could, saving the seeds for the seed company, building my plant collection for the eventual day when I could put down permanent roots somewhere. That eventual day was getting closer and closer. Both Jonathan and I hungered to plant fruit trees in a spot where we would still be around to eat them when they matured.

After three years at Wonder Bread, it was time for a change. I decided to let the seed company go. I had reached a place where I could take out a big loan and gamble on making the seed company take off or get a full-time job helping beginning farmers at the New England Small Farm Institute and start to have more financial stability, health care, and other grownup stuff. Although the seed company had a short life, it created the fertile ground for a new project—a book about perennial vegetables. I started writing *Perennial Vegetables* while *Edible Forest Gardens* was still under way. When we were looking for our next garden site, I was actively working on both books. The best parting gift from the seed company was Jonathan: he was here to stay. Many nights we were up late into the night researching strange crops on the Plants for a Future website, a database of seven thousand species of useful plants for the British climate. Jonathan was a stalwart, tending the garden while I was busy with my job, writing the two books, and in poor health. He grew into a plant geek of the highest order, a fantastic gardener, and a great friend.

SEEDS OF PARADISE
by Jonathan Bates

I'm always fascinated at how life unfolds. For me, it was how a fish and coral reef hobby became a career, how those interests morphed into a love of growing plants, and how that developed into a love for edible plants grown in gardens that mimic forest ecosystems. It was a wise man with knowledge of edible forest gardens who walked with me along the third unfolding.

Just before I left Vermont with my master's in social ecology, Eric Toensmeier, a favorite instructor of mine at the Institute for Social Ecology, approached me while I was doing research in the library.

"Jonathan, do you have any plans after you graduate?"

"Well," I said, "I haven't thought much about it."

"I'm looking for a housemate to garden with down in western Massachusetts. Would you like to come be an intern at my perennial vegetable seed company?"

Halloween night 2000 was a perfect time to visit Eric in Northampton. I remember the apprehension and excitement on an evening that I knew could head me in a new direction, down a new road in life. We dressed up in ridiculous costumes and went to a house party—music turned way up, dance floor pumping. After a few visits like that, getting to know the people and places of the Pioneer Valley, I took Eric up on his generous invitation and a great opportunity to meet other plant lovers, experience a new place, learn how to run a business, and plant a garden. By the beginning of the 2001 growing season, Eric and I were housemates and business collaborators.

Being a part of the seed company seemed simple enough: order bulk seed of some unusual vegetables; pour them into little manila envelopes; slap on labels; and sell each packet for $2.00. Money would be streaming through the door. Meanwhile, I would become a perennial vegetable expert!

Dozens of paper cuts and thousands of seed packets later, the romance faded and my perspective had changed. We hadn't yet planted our garden, so my only interaction with the plants was from the seed end. The diversity of sizes, colors, shapes of these seeds was interesting, but that initial fascination waned. We had bills to pay, customers to answer to, and lots of addresses to enter into spreadsheets. I learned a lot that first winter; it was an effective, low-cost way to learn about running a small business, despite the sore fingers. Most important, Eric and I became friends.

But, while our friendship grew, the business did not. Eventually Eric came to the conclusion that to make this venture a financial success, we would need not hundreds but thousands of customers to bring in enough income to pay one person's salary for a year. This realization was a huge blow. We were already spending long hours with our current customers; what would thousands look like? Soon both of us found other jobs. We stayed as housemates and started the garden at Wonder Bread Farm.

A kernel of Eric's wisdom stuck with me from that first year: "There may be a more viable business selling plants than seeds." I tucked that little seed of an idea in the envelope of my mind. Eventually, it would plant itself and begin to grow.

3

LET'S GET A PLACE

After gardening with him for a couple years, I knew Jonathan was a solid, friendly, reliable guy and that he and I shared an enthusiasm for our particular kind of gardening. I also knew that neither of us had any plans for the next decade or so. We were at a crossroads, and we bounced around various ideas about what would come next: go back to school, move to another region of the country, stay where we were.

Eventually, Jonathan and I decided we wanted to garden in the city. Two guys raised in the suburbs, active in urban agriculture and social justice couldn't commit to staying in the country forever. It may be counterintuitive, but city living is in many ways more ecological than rural life: for example, in a city you can walk to the library, farmers' market, and supermarket instead of having to drive everywhere. (Plus I hate driving on icy country roads all winter.) And even though we spent most of our time talking about plants, Jonathan and I did have other aspirations, like not staying single forever. Living in the country and identifying edible mushrooms in the backyard is not a good strategy for meeting women; we wanted to move where the people were.

Although I had closed down the seed company to take a job at the New England Small Farm Institute, it wasn't a move away from doing what I loved. It was a move toward it: part of why I got a real job was to start saving money and building credit toward homeownership so that we could get some fruit trees in the ground. Why keep waiting around in one rented place after another? I could feel those persimmons in my future.

While we were still in Southampton, Jonathan and I sat down and wrote a list of what we wanted in our next garden. Jonathan and I agreed to use the garden we were developing as a pilot test and case study for the design process Dave Jacke and I were putting together for *Edible Forest Gardens*. Permaculture design is about more than planning a garden. It starts with clarifying your goals, because if you don't know what you want, you are unlikely to achieve it. Before we found the site that was to become our home, we had already articulated most of our goals, and in fact they helped us determine what to look for.

Jonathan and I wanted to create "an intensively managed backyard foraging paradise": to take a walk through the garden every day and be able to pick fresh greens, fruit, and other foods, grazing on them straight from the plant. To be specific, we wanted to have two handfuls of fresh fruit every day for everybody in the house, including guests, for as long a season as possible. We knew that gardening in the city it was unlikely that we would have enough space to raise any individual crop in large quantities—to freeze enough blueberries to eat every day all winter, for example. That would have been lovely, too, and even as we established our goals for what we knew would be a small plot of land, we fantasized about a larger site where we could plant a huge blueberry patch or grow enough potatoes for a year or all our winter squash. But you can't always have everything all at once. We zeroed in on what was realistic in an urban lot and chose a long season of fresh eating as our primary driver.

As for choosing the species for our backyard Eden, we set out to create "a mega-diverse living ark of useful and multifunctional plants from our own bioregion and around the world." We wanted to paint our garden with a broad palette of species, emphasizing low-maintenance perennials that would provide fertility, weed and pest control, and tasty crops. A byproduct of this goal was a diversity of flavors and yields; we hoped (or perhaps joked) that we would grow at least three hundred species. This is probably not what most gardeners would set out to do, but for us experimenting to see what was possible was a big driver. If we didn't try out obscure species such as licorice milk vetch, hybrid arctic raspberries, and native perennial chickweed, who would?

Looking back, I think our goals are awfully fancy descriptions of things we now take for granted. But at the time they helped us clarify what we were up to and inspired us to keep going. We didn't want just one big edible forest garden; we wanted the perennial forest garden to surround other elements plugged into each other to create an eco-system. We wanted a warm microclimate for growing lush tropical food and foliage, intensive beds for annual crops like tomatoes and carrots, a greenhouse for winter salads, a pond and wetlands for growing aquatic vegetables, and perhaps small-scale fish raising. We also wanted social and functional spaces—a patio, a storage shed, pathways with partial pickup truck access, compost and mulch piles, even a shady, private "outdoor room" with perhaps a hammock.

We wanted our land to look and feel like a field in midsuccession, a mosaic of patches of shrubby thickets and small trees interspersed with more open spaces. We wanted some wild areas and some more manicured, to be able to try different effects and management styles. But overall the impression we aspired to was "ecofunctional," meaning intensively productive rather than ornamental.

And finally, as we began to look around at urban lots with an eye toward what we might buy, we wanted to be a bit like Dr. Frankenstein, to (as we wrote) "bring our dead and blighted backyard to life . . . creating a lush, semiprivate oasis that inspires our neighbors to plant their own." We wanted our garden "to serve as a refuge in our biologically impoverished neighborhood." It was too much to hope that moose and great blue herons would come to visit, but we did want to create a haven for smaller forms of wildlife, particularly those birds, insects, and amphibians that help control pests.

In writing *Edible Forest Gardens,* Dave and I pulled together disparate elements that we knew worked well on their own, drawing from the sciences of ecology and organic farming, gardening with native plants and unusual edible crops, indigenous land management, and tropical food forestry. As far as we knew, no one on this side of the Atlantic had ever put all those pieces together to create an integrated whole. Jonathan and I were excited to give them a try, to test what sounded great in theory and worked in Europe and the tropics. Although Jonathan and I had been able to trial many species at Wonder Bread, as well as test some nitrogen-fixers and nectary plants, we were not able to assemble them into an integrated whole; to come fully into their own, they needed to be grown in an agroecosystem with fruit trees, a pond, and more. And we hadn't truly been able to test out the challenge Dave and I were developing—to see if cold-climate forest garden systems were really viable.

Could a garden with trees, shrubs, vines, and herbaceous perennials produce food in every layer? Which low-maintenance perennial food plants were good to eat and productive, even in the shade? How did you cook with them? Would we have what it took to wait and let nature slowly respond to pest problems? Would we in fact have debilitating pest issues that our design approach could not handle? Would we introduce terrible new weeds? What if it was a disaster or a failure? What if there was something big we were forgetting? What if we planted a great garden and nobody wanted to visit it, learn from it, replicate it? These worries chewed at the back of our minds as we made our big plans.

4

GARDENING BEHIND THE TOFU CURTAIN

he Pioneer Valley of western Massachusetts has a deep socio-
economic divide. The northern counties are home to several
colleges and a large population of liberals and progressives. The southern part is more typical of the larger United States, including some
bigger cities with major challenges. Some folks who live in western
Massachusetts call the line between north and south the "Tofu Curtain." When we set out to find the perfect lot, Jonathan and I both
knew what side of the curtain we wanted to garden on.

That meant our forest garden would most likely be in the city of
Holyoke or Springfield. We wanted enough land to have a decent-
sized but not unmanageable garden. We also wanted an amount of
land close to that of an average American garden, so it would be
relevant to other people (if it worked).

Jonathan and I didn't know how long we would be spending there,
but we knew we would see it through for at least five years so we
could watch how the garden evolved—and taste the literal fruits of
our labor. We needed at least some areas with full sun, something
resembling soil on at least some of the garden, freedom from at least
massive populations of the worst weeds like Japanese knotweed or
poison ivy, and a livable residence.

During my twenties, I had been a part of many cooperative houses
and group living situations, so I knew they tended to be short-lived.
Jonathan and I both liked a lot of the benefits of cooperative living (such as low rent, shared cooking and chores, and having friends
around), but we wanted clear boundaries and long-term stability.

Once we planted that garden, we wanted to be able to live there comfortably, without getting in each other's way and fighting over the dishes.

How could we get the best of both worlds, sharing a garden without sharing a home? Part of our plan was to find sweethearts and entice them to move in with our edible backyard Eden. We were not naive enough to think that two women we had not even met yet were going to want to share a home just because their boyfriends were already there. My mom came up with the idea of finding a duplex with a shared garden. Jonathan and I could live together on one side and rent the other until our hypothetical sweethearts were ready to join us.

When we looked at a brand-new, modest duplex nearing completion of construction in a neighborhood of Holyoke we liked, we knew we had been right to pass on the handful of other possibilities our real estate agent had shown us. The house is at the top of a ridge in a mixed-income neighborhood of single-family homes, duplexes, and some large tenement buildings. From the second floor you can look down and see the Connecticut River in winter, and it is within walking distance of both a Kmart plaza and the farmers' market.

The moment we walked onto that steep front driveway, we started fantasizing about where the hardy bananas were going to grow and what a great tropical garden we could plant in all the heat that was rising up from the pavement. The backyard was big enough and full of potential. We were inching closer to putting our fancy-language goals and sophisticated theories to the test, on a real piece of ground—a compacted, bare, abandoned piece of ground.

Four hundred years ago, Holyoke's fertile floodplain was farmed by native people, probably the Pocomtuck Indians. When the English colonized the region, they also brought disease, sometimes intentionally, as with the "gift" to Indians of blankets from smallpox wards. Holyoke became one of the first planned industrial cities in the country, designed to make paper and other products, utilizing hydropower from a dam and an ingenious series of canals. Parallel but each lower in elevation than the last, the canals spill through factories to provide additional hydropower.

Waves of immigrants came to Holyoke to work in the factories—Irish, French Canadians, Poles, and more. Some local historians think factory owners recruited new immigrant groups to keep workers from being able to organize because of language barriers and cultural differences. In the 1950s, Puerto Ricans arrived just as the factory jobs began drying up. Holyoke is now part of what is known as the Massachusetts Rust Belt. It sometimes seems as if half the factories and tenements were burned down, leaving a landscape of empty lots and crumbling buildings. When I visited Detroit in 2009 to teach a workshop, I felt right at home.

In 2010 Holyoke was about 50 percent Latino, mostly Puerto Rican; the public schools were closer to 90 percent Latino. The town is on the verge of a big demographic shift. Somehow, however, until recently almost all city officials were white. Holyoke is one of the poorer cities in the country, and though small—the population is about forty thousand—it has typical inner-city problems like drugs, gangs, and unemployment. A few years ago the technical high school had a 37 percent graduation rate. Many neighborhoods have little access to fresh, healthy food. Obesity and diabetes are rampant. I was shocked to find out how many people in the city can't read. When I talk about Holyoke, people from outside the region often assume that must be where Mount Holyoke College is, and they imagine a quiet college town. Mount Holyoke College is a town over and a world away.

But the city of Holyoke is full of hope as well. Lots of people are trying to make things better. The affordability of real estate is driving some development, including a new wave of "immigrants" from some of the wealthier towns north of the Tofu Curtain. A company that converts diesel vehicles to run on used vegetable oil moved to town, and there are hopes for more green industry. Some of our neglected factory buildings are coming back into use, with businesses looking to capitalize on the infrastructure of the old mills.

Like some other postindustrial cities, Holyoke has experienced a renaissance in urban agriculture. Throughout the city, empty lots were turned into gardens and microfarms. Nuestras Raíces emerged in the early 1990s, establishing a network of community gardens, youth programs, and interlinked small businesses building on the heritage and skills of the city's Puerto Rican population. I was on the board of directors of Nuestras Raíces for nine years; just a few months after I joined the board, the organization hired a young director fresh off a stint as a migrant labor health worker in Immokalee, Florida. We had enough funds to pay him for six months and give him half an office the size of a closet. Over the next decade and a half, Daniel Ross would turn Nuestras Raíces into a million-dollar-a-year organization. He would also become one of my closest friends.

Jonathan had his own ties to the area. He worked for an urban youth gardening project called Gardening the Community in nearby Springfield. As we began looking for our new home, Holyoke became the obvious choice.

All of this is to say that permaculture design is not just about your yard. Understanding the broader social and ecological context is important to create a design that is as multifunctional as possible. Jonathan and I both felt that locating our garden in Holyoke would make it more relevant to the region (and the world's) urban population. Affordable home prices didn't hurt either. We moved into the duplex in January 2004.

REPRODUCING EDEN
by Jonathan Bates

While at Wonder Bread Organic Farm, I found that quiet rural life has a lot going for it: wildlife, sunsets, starry nights, solitude, hayfields, the smell of woodland after a summer rain. Yet without family or friends close by, fields and forests become lonely places. And while I was enamored with plants and in awe of the beauty, diversity, and complexity of our garden, I realized at some point that I needed to spend more time with other human beings. Maybe even find a partner to share my life with.

For this to happen, it was necessary to move closer to people. But where?

Using criteria that Eric and I came up with, in six months we found an affordable, urban, duplex with land. Although it was less land than we had hoped for—one-tenth of an acre versus two acres—it was enough to grow an awesome garden, and the brand new house was in a hundred-year-old neighborhood, close to public transportation and downtown, and within our price range.

To afford a home with our salaries in the economy such as it was during the early years of the new century, rental income from the other side of the duplex was necessary. In the meantime, like a bird builds its nest and lines it with shiny objects, we hoped to attract our future mates. Down the road, Eric and I would split the full costs, each of us living on either side with our families.

We were venturing into new territory with this idea. Neither Eric nor I had ever bought a house before. We were going against the convention of getting married first, then buying a house with a white picket fence, getting a dog, and having two and a half kids. We had no examples to look to—two single friends buying a house with the intent to divide the investment equally, sharing the debt, land, and everything that comes with that together as two future families. Trusting each other with such a responsibility felt especially rare in this world.

While Eric and I fantasized about the possibility of finding a house in the city to plant our garden, my parents were asking important questions like, "What kind of mortgage can you afford?" And, "Doesn't a bank want to see stable income and good credit?" It was this kind of informed thinking and the excitement of not just helping us but also looking at this as an investment for themselves that led my parents to generously purchase the house for us and rent it to us for several years until we were ready to buy it ourselves. They were the original gardeners who fertilized the project, allowing the seeds to grow.

Every once in a while I think back on the process we went through to find our home in Holyoke. I also reflect on other points in life that have shaped me. For example, I was born on a farm commune in the country, a brief but influential experiment my parents experienced. Might my story of living with friends collectively in an urban garden really go back to that Virginia commune? Of course starting a family and creating a paradise to call my own can't be complete without an Eve. . .

5

SUN, SHADE, SOIL, SLOPE

Permaculture design is about harmonizing your goals with the unique characteristics of the site. Rather than laser leveling and planting uniform rows, permaculturists strive to understand what's happening on a piece of land: What is the overall pattern? What are the opportunities and limitations? What potential can be unlocked through regenerative land use? At first glance our backyard looked like a blank canvas—a raw slab of bad soil, uniform and monotonous. But as we studied the site, we discovered a different, far more complex (and complete) picture.

Sounds great. In order to get from here to there, however, it meant drawing a fairly tedious but truly worthwhile series of maps on tracing paper. Each map viewed the site through a different lens—soils, sun and shade, slope—each with an effect on a garden's ability to grow and thrive. As we made the maps, we were observing, studying, and talking about each aspect, each of which presented us with a different vantage point from which to see what was happening on the land and a degree of focus that deepened our understanding. At the end of the process, we laid multiple layers of tracing paper on top of our base map to see a combined image illuminating our site's less obvious secrets.

This phase of the permaculture design process is known as analysis and assessment. Analysis is about observing current conditions on a piece of land. In theory it does not presume to like or dislike, only to describe objectively. In practice this can be a difficult Zen exercise; an observation like "Norway maples cover much of the backyard" can quickly lead to feelings of frustration and discouragement. However, in the assessment you interpret the results of analysis through the lens of your goals for land use and evaluate the situation. For Jonathan and me, distancing ourselves like this helped, and we began to find peace with our towering Norwegian neighbors.

Permaculture experts say that you should spend at least a full year observing your site before deciding what to do with it. I had heard this many times but until now had never had the chance to try it for myself. We moved in in January 2004 and spent the year on observation and design. Along with drawing the maps, seeing the land in all its seasons gave us essential insights, not least of which was understanding the seasonal patterns of sun and shade, which provided the key insight that our design crystallized around.

Our future garden had some unusual patterns of light and shade, with different areas having distinct characteristics. The front yard faced southeast, and we could see immediately that it got great morning and early afternoon sun. This area was also protected by the house from cold winter winds, making it a warm, sheltered microclimate. Later we would find that this part of the garden had a longer growing season by a week or two on either end. In fact, much to our surprise, we were able to overwinter some plants rated at USDA zone 8 (hardy to 10°F), though we are rated at zone 6 (hardy to 10°F).

The alleys to either side of the garden had close to full shade year-round. On the south side this was courtesy of a twenty-five-foot arborvitae hedge that eliminated the possibility of sunny windows or a south-facing attached greenhouse. To the north the house itself provided heavy shade for most of the day almost all year. The backyard had an interesting pattern: the neighbors' barn-sized garage on the south property edge cast heavy shade in winter on the southern half of the backyard when the sun angle was low. In summer the same area of our garden received full sun for at least six hours a day. The north side of our garden had the opposite situation. The winter sun's low angle cut below our northern neighbor's overhanging Norway maples, providing full sun in winter. But by midsummer the same area was in close to full shade from the maples.

We used some high- and low-tech strategies to determine sun and shade patterns. Jonathan took photographs out of the second-story window at different times of day and in different seasons. Dave Jacke came over with a handheld solar-finding device, which backed up Jonathan's observations. (Of course today they have an app for that.)

Because of the high density of buildings and trees in our urban neighborhood, only one tiny area of the garden, at the far end of the backyard, had sunlight in both summer and winter. This insight, not obvious at first glance, was the seed from which our design grew. To meet our goals we needed a greenhouse, and the site provided us with the welcome constraint of only one possible location. The first colors had been dabbed on that blank canvas, and the rest of what we wanted to paint had to work around that.

It was quickly apparent that we had several distinct types of soil. The owners before us grew marijuana in the basement; when they went away for vacation they forgot to set the timer on their grow lights and burned the house to the ground. When we moved in, the new house had just been completed, including excavation for the new basement. The half of the yard directly behind the house was fresh fill; it was mostly bare and heavily compacted by construction equipment. The main ingredient was clay subsoil, but it also contained chunks of concrete, brick fragments, and pieces of asphalt. Together with rebar and plastic, these form a new mineral class that city gardeners call "urbanite." So drainage was going to be a challenge, organic matter was absent, and we would have to address compaction. Mineral nutrient levels were low (phosphorus) to okay (potassium and calcium). The good news was that our soil tests indicated that the pH here was close to neutral, and there was no lead problem.

The other half of the backyard was made up of the remaining soil from the previous yard and is in some respects more typical of the oak forests that once covered the area. It was sandy and acid, with okay (potassium, phosphorus) to poor (calcium) mineral nutrients. There was at least some topsoil to speak of, which was a bit deeper under the Norway maples. It supported a somewhat weak growth of crabgrass, with some other urban weeds like ragweed and a feeble patch of Japanese knotweed. In *Wild Urban Plants of the Northeast,* Peter Del Tredici says of crabgrass (our dominant species): it "is remarkably drought tolerant and is common in trampled lawns in minimally maintained public parks and residential landscapes, unmowed highway banks and median strips, small pavement openings, and sidewalk cracks"; it is a "disturbance-adapted colonizer of bare ground; tolerant of contaminated and compacted soil." Not exactly a ringing endorsement of our yard's health. The fence lines featured typical urban, spontaneous vegetation like Norway maple and bittersweet, along with tattered forsythia and some feral red raspberries.

Our soil test also told us that the soil in this part of the backyard had lead—454 parts per million, which the extension office at the University of Massachusetts considers a "low" amount. Jonathan and I were prepared for some lead, as it's common in the Northeast's older cities. If it had been worse, it would have been a major factor in our design. We would have had to cluster edible greens in raised beds or in our compacted soil areas and focus on fruits (which are lead-free, unlike roots and leaves) and support plants in the back half of the garden. As it was, the results of our soil test recommended we bind up the lead by increasing the pH, improving organic matter, minimizing tillage, and mulching—all things we were going to do regardless. The danger from dust in low-lead soils is greater than the danger from eating foods grown in them.

When we asked the neighbors about the history of the site, we learned several possible explanations for our lead problem. First, there used to be apple trees in the backyard. In the early 1900s, lead arsenate (lead with arsenic, a great idea) was a popular pesticide for apples.

Rounding out our trio of terrible soils were the side alleys and the tiny strips of soil surrounding the driveway. These were a patchy blend of sand and gravel fill with some low-grade topsoil in a few areas. Mineral nutrient content was low for most of the major and minor plant nutrients UMass tests for. At least the north alley had some vegetation: a few legacy ornamentals and a large tree of heaven.

Readers in drier areas where xeriscaping and other water-conserving garden techniques are more common may be surprised that we did not pay more attention to water. In much of the world, water is a critical limiting factor for any kind of agriculture. But our climate provides forty inches of rain a year, broken down fairly evenly month by month. During the twenty years I've lived in Massachusetts, I have rarely seen a month go by without at least some rain. In July and August we can have a few hot, dry weeks, but they are usually followed by torrential thunderstorms.

Jonathan and I had hoped to capture rainwater and use it to irrigate our gardens. At Wonder Bread our house had been at the top of a ridge, with the gardens down below. You could have filled a cistern or rain barrel near the house, run some drip irrigation through the garden, and let gravity do the work. Unfortunately, our new home had the wrong slope for this. The backyard slanted gently up from the house, the incline mild enough that we felt at the time that rainwater-harvesting earthworks like contour swales were not worth the effort and constraints they would place on our design. In retrospect, we wish we had given this more thought.

We identified issues other than water conservation and rainwater harvesting that would affect the success and productivity of our design. The first was aesthetic. Our backyard was barren and exposed. It seemed as if every house on the block could see everything that was going on there. Then again, there was no reason for us to go back there, nothing to draw our interest or even anything that needed doing. Northwest winds swept right into our backyard down a canyon formed by a row of houses.

Using a process Dave had developed for *Edible Forest Gardens,* we also studied our ecological neighborhood. Our job was to evaluate food, water, shelter, and other factors that impact populations of beneficial insects and birds that eat pest insects. Our neighborhood and yard scored poorly: no open water, few nectar sources for insects, and little plant food for birds. We also lacked the "lumpy texture," or diversity of habitats and successional stages, that provide a healthy home for our pest-control allies like beneficial insects and pest-eating birds. There was some cover and habitat in the form of evergreen hedges and thicket-forming shrubs, as well as a narrow band of woods on the cliff across the street, though it had poor diversity, little understory, and none of the lumpy texture we want to see. Our overall rating was fair—one step up from a Walmart parking lot.

Site analysis also includes a look at legal restrictions (that is, what can you pull off?). Massachusetts is one of the most highly regulated states in the country, and Holyoke was no exception. Many sustainable practices, essential for the long-term survival of humanity, are currently illegal. So there were to be no simple composting toilets (even a high-tech, $5,000 model would have involved a drawn-out legal battle), no treatment and use of greywater in the garden (though this would reduce pressure on the city's overloaded combined sewer and stormwater system), and no livestock except rabbits, which are considered pets.

Given the legal landscape, Jonathan and I decided to choose our battles. Because we wanted to trumpet our garden to the world by inviting people for tours, there were certain things we would have loved to do that simply weren't going to happen. We were especially sad to close the door (so we thought) on chickens, as they were explicitly prohibited.

After our series of investigations into the property, we laid our single-issue tracing-paper sketches over our base map to see what the combined picture looked like. Sun and soil were clearly the most important factors around which we would build our design. The overlap of soil and light created a mosaic of patches with different conditions. We created a summary of our site analysis and assessment efforts that identified each of these distinct patches, characterizing their conditions, challenges, and limitations. We also noted our initial ideas about the suitability of different patches for our desired garden elements like compost areas, a shed, a pond, and vegetable beds.

Our front yard was a no-brainer. The moment we walked onto the property, we knew the front yard would be perfect for a tropical garden. Our site analysis confirmed that the southeast-facing, steep asphalt driveway, combined with the house, which would block cold winds, would create an island of Washington, D.C.–like climate. Although the soils were virtually sterile fill, this area had a lot to offer.

The alleys to the north and south of the house were both shady with bad soils. We knew we were going to have to use one of them as vehicle access so that we could bring building and gardening materials into the backyard, but we were not yet sure which one.

Our compacted-fill zone had several distinct patches. On the extreme north and south edges there were areas with full shade in summer; the shade, combined with terrible soil, meant these edges were particularly poorly suited for growing food. It was going to take some creativity to find a productive use for these zones.

The sunny parts of the compacted-fill zone, which represented at least half of our land with full sun, presented a more interesting opportunity. With work, we could improve the soil, but that area might be more suitable to a pond or raised beds. Further, our house had two back doors (one from each side of the duplex), and both opened into this area; logically, this would become our main social space. The house rose up from the backyard like a featureless cliff, and we wanted to visually anchor it to the garden.

In our areas with "good" soil (which meant that it was sandy, with some lead), we again had three divisions based on sunlight. The only spot for a greenhouse was right in the middle—the sweet spot that would have sun all year. Our sandy, acid, lead soil zone also had both sunny and shady areas in summer.

Now we knew some of what our site had to offer. We had spent a year observing it in all its seasons, analyzing its various aspects. We learned that it was far from the uniform empty lot it first appeared to be but rather a patchwork of islands with different conditions and different potential for helping us realize our goals. Its challenges were also apparent: shade, lead, compaction, poor aesthetics, and minimal habitat for birds and beneficial insects. But by understanding the challenges and some of the aspects that were never going to change, we began to see the possibilities as well.

What we needed next was an overarching vision to tie it all together, a layout that would harmonize with the patterns of the landscape. It was time for our first real experience with design—and to make choices we would have to live with for as long as we were there.

6

A MODEL ECOSYSTEM . . . BEHIND KMART

The kinds of plants that grow in neglected urban wastelands don't get much respect. But they actually have essential roles in ecosystems: healing degraded soils, cooling summer temperatures, sequestering carbon, controlling erosion, providing food and habitat for urban wildlife, and helping rainwater infiltrate the soil rather than run off. In *Wild Urban Plants of the Northeast*, Del Tredici calls these plant communities "spontaneous urban vegetation" and champions the important contributions of maligned plants like tree of heaven, Norway maple, and crabgrass.

Del Tredici's definition of "urban" is anywhere pavement or buildings cover more than half the soil and essentially nothing remains of the original preconstruction ecosystem. The neighborhood we moved into was about at that 50 percent mark; the only remnants of the original ecosystems were on a steep, trash-strewn cliff across the street from our house. Ecologists might call our neighborhood "depauperate," defined as a poor habitat due to lack of species diversity. In other words, we had just moved into a typical Northeastern neighborhood.

Wild Urban Plants of the Northeast describes some of the typical disturbances urban plants must be able to tolerate. These include extensive paving, which besides being a bad growing medium can cause drainage problems (both flooding and drought); compacted soils; heavy use of road salt, which makes soil alkaline and toxic to many plants; and of course contamination of air and soil. These stresses make plants vulnerable to pest and disease pressure. Plants that can withstand these adverse conditions are members of a tough and resilient club.

Del Tredici says that ecological restoration with native plants in cities is just gardening, as you will forever be weeding out robust members of the new, spontaneous urban vegetation of the city as they return to their "natural" habitat. Some species are originally from the Northeast, others are from Eurasia or elsewhere, but all are adapted to the terrible disturbances and awful growing conditions that we have created in our cities.

The more I learn about ecology and permaculture, the more I have come to appreciate what dense tangles of tree of heaven, mulberry, sumac, and goldenrod are doing: healing cities from the damage we have done to them. In Europe some abandoned lots that have grown up into spontaneous urban forest or woodland have been made ecological preserves to honor the important role these emerging eco-systems play. Because a core concept of permaculture is using nature as the basis for design, Jonathan and I knew we would need to look to nature and, more specifically, maligned nuisance plants to inform and round out our vision.

As Dave and I were writing *Edible Forest Gardens,* one of our goals was to provide a detailed basis for ecosystem mimicry. We spent years reading about different ecological communities within the eastern forest region (between the midwestern prairies and the Atlantic Ocean, a broad zone of forests) and visited some of the tiny remaining fragments of old growth. I've explored reasonably healthy pine barrens, coastal scrub, oak-hickory, old-growth northern hardwoods, and other ecosystems. Just as these may contain habitats like forest edge, wetland, and hot, dry south-facing slopes, urban ecosystems have their habitats as well. Chain-link fences, parking lots, and side-walk cracks provide niches for plants and opportunities to build soil and bring an irrepressible urban ecosystem to life. An old-growth forest with deep, healthy soils and a web of intact relationships was a poor model for Jonathan and me. We needed to look toward an ecosystem that reflected our sterile fill and lack of biodiversity.

After we moved in, Jonathan went on regular hikes around the neighborhood to get to know the area. One day he found a little paradise behind a run-down Kmart plaza about a mile southwest of our house. Right away he took me to see it.

I have always loved thickets, old fields, young woodlands, and other midsuccession habitats, which tend to have sun but not full sun and, it seems, the best wild edibles. This feral ecosystem behind Kmart featured over ten acres of dense clumps of thicket-forming shrubs interspersed with open wildflower meadows. Here and there some teenaged oak and black locust trees emerged. Songbirds and sparrows were everywhere, and once you walked in a few feet, it was easy to forget you were close to roads, shopping centers, and housing projects. Bumblebees and other pollinators buzzed around a myriad of flowers, from thistles to black-eyed Susans.

When the Kmart shopping center was put in, huge amounts of soil were moved around by heavy machinery in order to level and excavate the parking lot and structures. What remained was an infertile and compacted soil—much like our own backyard. But because Kmart had been built about two decades earlier, an ecosystem was emerging that we had a lot to learn from; it was a window into what our own soils and vegetation might look like in twenty years.

There are no plant communities native to Massachusetts that grow on Kmart construction fill. But some native and some newly arrived species had managed to find a way to grow and even thrive there, forming a new hybrid ecosystem. Multiflora rose and bittersweet were prominent, along with underappreciated natives like sumac and goldenrod (both valued as ornamentals in Europe). Some of the soils were so poor that the only trees for acres were nitrogen-fixers: black locust and autumn olive.

Most gardeners would not be excited about the species that were growing in the abandoned area behind the shopping center. But to me, any plant community that can grow in such terrible conditions is a welcome one. Jonathan and I wanted to adopt this lumpy patchwork of shrub and meadow communities as the model for our own garden.

That's not to say we wanted to plant so-called invasives like bittersweet or multiflora rose in our garden, nor did we want aggressive natives like sumac or poison ivy. But we wanted to mimic the mixed texture, the patchwork of shrubs and meadow with scattered trees. And we particularly wanted to mimic the process by which terrible post-construction soil was converted into a productive, living ecosystem.

By choosing our particular home, Jonathan and I were following the principle of site repair. The essence of site repair is that you don't build your home on an old-growth forest; instead, you take responsibility for places that have already been extensively disturbed by humans and heal them as you garden and build. Lucky for us, there was hardly any way we could have made conditions in our garden any worse.

This is an example of the process of regenerative design, which asks us how our designs can bring a site to life and bring us into a deeper relationship with it and each other through doing so. While sustainability is focused on maintaining things as they are, regenerative land use actively improves and heals a site and its ecosystems. Regenerative agriculture, which permaculture aspires to be and often actually pulls off, achieves these goals while also meeting human needs. It's kind of an important topic for humanity this century.

7

GUILD-BUILD

One of my favorite phases of any design is assembling a species palette, a master list of all the species you might use to paint a living and productive landscape onto your site. The "guild-build" process that Dave and I developed for *Edible Forest Gardens* helps gardeners assemble a master list of species for all necessary niches.

The first phase of guild-build is to make a list of all the things you hope to grow. The woody plants Jonathan and I were most excited about were American persimmon, pawpaw, chestnuts, Asian pear, and hardy kiwifruit. We already grew most of the herbaceous species that we were keen on, from good King Henry to strawberries and perennial ground-cherries.

After you have analyzed your site, you look at your list of desired species to see what's realistic and what's not. Sadly, some species are usually cut at this point. For us, the list was long and, particularly with the trees, impossible. There simply would not be space for chestnuts: the two full-size trees required for pollination would fill up almost the entire yard. We also regretfully closed the door on nut pines, the macadamia-like nuts of yellowhorn, and figlike che fruits. In fact, given the small size of our lot, we were going to have to work hard to achieve our goal of a double handful of fresh fruit every day from May to October. To do that, we would have to focus primarily on fruiting shrubs and dwarf trees. Besides space, our other primary limitation was shade. We were going to be able to grow only a few species that required full sun; we had more room to stretch out and explore the range of shade-loving edibles.

Another part of the guild-build process is determining what uses and functions are called for and cross-checking it against the list of everything you hope to grow. We were going to need groundcovers, nitrogen-fixing species, and insect nectary plants. Jonathan and I looked for gaps in our list based on the roles we wanted our plants to serve in the garden. Did we have nitrogen-fixing species for shade? Had we included any evergreen groundcovers? We went back to the tables in *Edible Forest Gardens* (still unpublished at this point) to select species to fill the niches that we had left open.

Jonathan and I were guided by the principle that everything we planted should have multiple functions and should be edible whenever possible. We also wanted to begin our search with native species and expand outward from there. The challenge was that maximizing diversity was also a priority for us; we wanted to sample the range of possibilities, especially for smaller plants like herbaceous perennials.

Very few of the foods Americans eat on a regular basis are native; the only ones you can really buy at the supermarket are pecans, sunflowers, wild rice, blueberries, and cranberries. On the other hand, there are many native species that have potential but are still minor foods; they are simply not domesticated to the point that they are ready for prime time.

Jonathan and I knew there would be core native edibles in our garden that would serve as anchors in our species palette. The native fruits we chose to include were American persimmon, pawpaw, beach plum, clove currant, various blueberry species and hybrids, multiple juneberries, and many more—a total of twenty native fruit species in fifteen genera. Not bad for a tenth of an acre. We set out for about half of our garden to be natives, which would mean a hundred or more representatives of the eastern flora. Most nut trees were too large for our garden, but we tracked down two chinquapins (native bush chestnuts) from a small nursery in Georgia. And we included many species of native herbaceous wild edibles, from sunchokes to giant Solomon's seal. Though none were as far along in domestication and productivity as a perennial like asparagus, we felt it was important to include them. From nearby forests we collected seeds of native nitrogen-fixers, like tick trefoils, hog peanuts, and wild sennas, and nectary plants, like the impressive cow parsnip. Collecting seed of wild plants is fun and, as long as the plants are not rare and you leave plenty of seed, nothing to feel bad about. In fact, by taking the plants into cultivation, I feel we are reducing pressure on wild stands. I'd been exploring the areas behind bowling alleys and beneath highway overpasses for a decade and knew right where to watch while driving by. When I saw seed drying, we pulled over (perhaps sometimes recklessly) and added to our collection.

At the 2010 annual conference of the Ecological Landscaping Association in Springfield, Massachusetts, there was a panel discussion, which quickly turned into a debate, about the role of nonnative species in ecological landscaping. When I commented to the panel that native plants are not the only "ecological" landscaping issue, I was surprised that some people in the audience reacted as though I had said something scandalous. To my mind the important question is not solely whether or not your garden is made up of all native species but where your food comes from. An all-native garden can, for example, provide an excellent habitat for wildlife, so you are clearly doing something right, but the challenge is that very few native species have ever been properly domesticated, so if you grow native plants, but your food is grown on industrial farms in China or Chile and shipped halfway around the world, you still have an enormous ecological footprint. My current approach with native-plant enthusiasts is to encourage them to grow, eat, and eventually domesticate more native species. It's an interesting challenge, and many of them are responding positively. Meanwhile I'll continue to grow and eat pears and asparagus alongside my native persimmons and black nightshade. Our home gardens can provide both habitat *and* food, and the place to begin is by familiarizing ourselves with edible native plants.

For Jonathan and me, considering the larger context of fossil-fuel use, the industrial food system, and climate change, we chose to cast a wide net for our garden's inhabitants and have not regretted it. Though we started off with natives, we also wanted a maximum of edible and multifunctional species, and that meant we needed to sample broadly, regardless of origin, because the native species palette we came up with left us with a lot of gaps to fill.

Our search for multifunctional species led us down some unexpected paths. For example, we didn't feel we could sacrifice much space to nitrogen-fixing plants that were not also providing food. This led us to hog peanuts, a shade-loving native with edible underground beans. They are quite lovely and often seen alongside the path on New England hikes, but except at Tripple Brook Farm I had never seen hog peanuts growing in anyone's garden. It is the perfect example of an underappreciated native species, which gains importance in a garden that prioritizes multifunctional plants to fill specific niches.

Not all of our choices were as easy. For example, there is no eastern native nitrogen-fixing shrub with decent edible fruit. We asked ourselves what we should pick instead: a native nonedible like sweetfern, a nonnative edible like goumi, or even a nonnative nonedible like Siberian pea shrub? Given our space constraints, we went with goumi, a relative of the maligned autumn olive, because it both fruits and makes fertilizer, which no Massachusetts native species could do. This medium-sized Asian shrub is a great nitrogen-fixer and grows well in the Northeastern United States.

Up until we completed the species palette, Jonathan and I did not need to show much originality in our design. Our goals were our own, but the final analysis and assessment map we created, though it included a few suggestions about what might happen where, was basically a final report to ourselves about our yard. We had selected a species palette and thought about what kinds of habitats we wanted to emulate but had no formal layout for where everything would go and what habitats we'd develop where. It was time to find the overall pattern that could be laid across our battered backyard to regenerate life and productivity. In other words, we needed to develop our design concept: a basic sketch and an elevator speech or sound byte that would capture the essence of the design.

Jonathan and I played around with several different design concepts. One leapt out as the best way to respond to the site's challenges and realize our goals. While in the unpacking we changed some of those details, when I look back on our original concept, I see the essence of what our garden is today: "Fingers of annual vegetable beds reach south from the greenhouse and the tall, shady forest garden, interlacing with fingers of low, sun-loving trees and shrubs. Near the house, social and functional spaces revolve around a cluster of three water gardens."

We decided to use the south alley for our access road: both alleys had equal shade and width, but the southern alley would do a better job draining frost from our garden.

Its bad soil and full shade made it ideal for offloading and storing piles of compost and mulch. The north side of the garden, with summer shade, would become our woodland edge habitat, with shade-loving species like pawpaws, gooseberries, and wild leeks growing under Norway maples. We would build a shed in the area with terrible clay soil and summer shade on the north side of the property.

We already knew where our greenhouse had to go: in the small year-round sunny spot between the summer sun and winter sun areas, and we decided to lay our main path leading from the house to the greenhouse along the line between the summer sun and summer shade areas. The summer sun areas would mimic the old-field mosaic habitat behind Kmart and feature shrub and perennial beds alternating with annual beds.

It was clear that Jonathan and I had found the challenge we'd been looking for. Could we bring about an edible paradise on our blighted lot? Could we regenerate soil, bring back birds, and meet all of our goals on only a tenth of an acre without cramming everything in too tight? And might we ever meet women who could appreciate guys who spent more time on the Plants for a Future online database than singles websites? Time would tell.

PART TWO

CREEP

2004–2007

8

IT TAKES A VILLAGE TO PLANT A FOOD FOREST

Jonathan and I wanted to spend a full year getting to know our Paradise Lot before we made final design decisions; ideally we would not have done anything on the land during that time. However, although the design process sounds rigid and linear, it actually loops back on itself like a bowl of spaghetti, each element informing all the others. So we allowed ourselves to do a modest amount of work on the land and let that influence the design still in process. Plus we simply had some things we couldn't wait to accomplish. Most important, we had a lot of perennials still over at Wonder Bread that we wanted to bring to their new home. And there was no question, no matter how the design finally ended up, that we were going to need to build our soil; that first year we figured we might as well get started.

When we moved into the house in January 2004, several hundred perennials remained in the frozen ground under snow at Wonder Bread Organic Farm. So that April we put out a call for a weekend work party on our western Massachusetts permaculture e-mail group. At least fifteen friends, coworkers, and strangers showed up each day, tools and gloves in hand, to help. In the permaculture community this kind of modern barn-raising, known as a "permablitz," can rapidly transform a landscape.

On the first day of our work party, our goal was to sheet mulch a thirty-by-fifty-foot nursery in our new garden that could hold the perennials we planned to bring over from Wonder Bread. Sheet mulching is a way to build soil instantaneously and without till-ing. The technique is sometimes referred to as "lasagna gardening" because it involves adding one layer of organic material after another, like making a lasagna. It's also known as "sheet composting" because you essentially build a compost pile spread out over the soil. Whatever you call it, this is a fast and versatile way of gardening, especially in cities, because it allows you to grow on top of a lawn or asphalt (or even a roof) without having to dig anything up; you just pile things on. When we lived at Wonder Bread Farm, Jonathan and I sheet mulched on top of the lawn, producing heavy crops of tomatoes and sweet corn the first year.

Sheet mulching has several important components, each constitut-ing its own layer. The first layer is made up of soil amendments like lime and mineral fertilizers. Jonathan and I spread lime and greensand in accordance with recommendations from our soil test because some of our garden was acid and all of it was low in micronutrients.

Along with the soil amendments, you can add organic material that might have weed seeds or propagules, like hay or manure. You can pile this organic material pretty high—six to eighteen inches. We did not have a lot of organic material since nothing much was growing in our yard yet, so we mostly just threw down the amendments.

The next layer is a biodegradable weed barrier, which, when built on top of a lawn, smothers out the grass or weeds until they break down and become worm food. This is why you can include organic material that might have weed seeds or propagules as your first layer; under the weed barrier, those seeds won't germinate and usually decompose or are eaten by invertebrates by the time the biodegrad-able weed barrier rots away and breaks down into compost. This takes three to six months, depending on the weather, the time of year, and the available materials.

Our weed barrier of choice is cardboard, preferably large pieces because they are easier to handle and thus faster to lay down than newspaper or small pieces of cardboard. They also leave fewer potential gaps where weeds can poke through. For our weed barrier,

Jonathan headed out one spring morning in his 1981 VW Rabbit biodiesel truck and stopped at appliance stores and bike shops, the best places to find sizable cardboard boxes.

On top of the weed barrier, you add layers of weed-free, carbon- and nitrogen-rich materials, again like making a compost pile. Carbon-rich materials are typically dry organic matter like straw, sawdust, or dead leaves. High-nitrogen materials include food scraps, manure, and fresh green leaves or grass clippings.

Jonathan and I took advantage of living in the city by collecting about seventy bags of leaves that people put out on the curb for pickup. Those leaves became the basis of our soil in much of the yard. Our main nitrogen source the first year was fresh-cut grass clippings, though we later discovered that it wasn't enough when we started to see some yellowing and slow growth on many of our plants. Household liquid activator (urine diluted five to one with water) contains a lot of nitrogen and did the trick at an affordable price. We also made the mistake of using whole leaves instead of shredding them first; when it rained, they turned into sticky layers of wet leaves that did not break down or breathe well. (The following year we purchased a leaf shredder to speed up the decomposition process and prevent wet leaves from matting, but it was temperamental and only worked on extremely dry leaves.)

Our neighbors and workday participants were amazed by the sheet mulch technique. The experience of turning a lawn into an instant garden using free material that most people consider to be waste has a lot of power.

On the second day of our permablitz, we all drove over to Wonder Bread, dug up the two to three hundred plants that Jonathan and I had grown from seed, and brought them to their new home. It was quite a project to dig up the plants; pack them into pots, milk crates, and recycling boxes; and load them into the trunks of cars, the backs of pickups, and one small trailer that our friend Jono Neiger had brought. But all the work was worth it. We hauled good King Henry, sunchokes, and groundnuts; nectary plants like sweet cicely and anise hyssop; nitrogen-fixing groundcovers like licorice milk vetch and lupines; and groundcovers including strawberries and violets we dug up from the Wonder Bread parking lot.

When we held these work parties, Jonathan and I were so grateful for the help that we tried to feed people well and send them home with plants, seeds, and produce. It didn't take long for us to notice, however, that we weren't the only ones who were benefiting from those workdays. Many people showed up excited to be part of an innovative project, learn new skills, and experience the joy of starting a garden. Jonathan and I both feel a commitment to give back to the community, partly by involving others, making our garden the best it can be, and hosting tours. Those early days of building the soil began to build our community, too.

Even though planting the nursery was several steps ahead of where we were supposed to be in the design process, as we analyzed the site and worked on our design over that first year, the things Jonathan and I learned in the nursery influenced our thinking about the site. We gained a better understanding of our so-called full sun area where the nursery was placed, discovering it had only eight hours of proper full sun for a few weeks in June. We also had a chance to test our sheet mulch technique before applying it to the rest of the garden.

We didn't stop sheet mulching when the nursery was completed. There was no scenario in which we were not going to sheet mulch the entire yard, so as labor and materials permitted, we kept going through the rest of 2004, starting with the front yard, which was to become the tropical garden. Sheet mulching is a labor-intensive process, so we did about a third of the yard each year from 2004 to 2006. As Jonathan and I would come to realize, sheet mulching didn't resolve our soil's underlying compaction in the central part of the yard, but it did enable us to transform our crabgrass wasteland into the beginnings of our foraging paradise.

9

TACKY TROPICALESQUE TAKES OFF

Jonathan and I wanted a showy, even gaudy, tropical welcome mat to introduce ourselves to our neighbors, half of whom or more grew up in the Caribbean, so we couldn't have asked for anything more than the warm microclimate of the front yard.

I fell in in love with tropical landscapes—characterized by big, bold leaves; bright foliage; and exotic flowers and fruits—at Florida's Selby and Kanapaha Botanical Gardens. And I was inspired to think I could do it myself in Massachusetts (only more edible) from reading books about cold-climate tropical gardening like Susan Roth and Dennis Schrader's *Hot Plants for Cool Climates* and David Francko's *Palms Won't Grow Here.* Although Jonathan and I chose some tropical-looking perennials, notably hardy bananas and passionfruits, as anchors, we knew this was one place we were going to be trying new heat-loving edibles, like edible hibiscus and taro, every year. Some would be true annuals even in the tropics, others perennial there but not for us. A single application of sheet mulch was enough to get us started.

The centerpiece of the front yard was to be a pair of hardy Japanese fiber bananas. These relatives of edible bananas don't make edible fruit. In fact, here in Massachusetts they don't even have time to flower, but they overwinter reliably. Even though we were otherwise pretty militant about not growing plants that are purely ornamental, we chose to grow them anyway, because they make such an over-the-top statement of fecundity.

Beneath the bananas we planted a number of perennial and self-sowing vines and groundcovers. We planned for a seasonal sequence of emergence, which turned out better than we deserved. The first plant to come up in the spring is chameleon plant, a multicolored, variegated groundcover with leaves that taste like ginger and fish; in small amounts these leaves add flavor to coconut curry. For many gardeners, chameleon plant becomes a serious weed, but we knew it would be swamped by vines to the point of invisibility by midsummer even the first year, so we weren't concerned about keeping it under control.

The next to appear is Chinese yam, a vine with cinnamon-scented flowers that in the fall produces hundreds of pea- to chickpea-sized tubers on the vine like little fruits.

Around late May, seedlings emerge of cucumber berry, an obscure native species that makes a great groundcover and produces crisp and lemony half-inch cucumbers (though the plant climbs a bit much). Finally, in early June, maypop shoots come flying out of the soil. This native passionfruit is a spectacular ornamental with intricate flowers that bumblebees rest on at night. Maypop is used also as a chamomile-like calming tea.

A fig tree grows between our bananas. Jonathan's parents gave us the Chicago Hardy fig variety, which dies back to the ground in winter but, unlike most figs, fruits on first-year wood. So when it resprouts eight feet high each year, we get a handful of fresh figs. Not bad for Massachusetts!

We decided to purchase or grow our own tropical annuals each year for the strip between the driveways, although most years tall, tender perennials came to dominate this area. Moringa, a highly nutritious tree with edible leaves, grew an incredible eleven feet tall and even flowered in a single season from seed when started indoors. Chipilín, a favorite Central American leaf crop, grew an impressive nine feet high and was covered in yellow, pealike flowers. We have grown several classes of cannas and experimented with cooking their edible tubers, though the best edible cannas don't have a long enough season to yield well in Massachusetts.

My favorite bold tropical is cranberry hibiscus. This shrub grows eight feet high and has burgundy foliage that looks like Japanese maple. The leaves are sour and fun to snack on right in the driveway. They are at their best cooked, though it took us some time to find the best preparation. They dye chicken meat a sickly gray, but they turn eggs a neon magenta, and in a frittata, with some onions and kale or other leafy greens, they are sublime. The best midsized species we've grown between our asphalt pads is eggplant. I think slender-fruited Asian types like Ping Tung Long are the best for eating, and the deep purple stems and lavender flowers are a nice ornamental as well. We have had good luck with Redbor kale, which, though not tropical, has a dark purple color and palmlike appearance that makes it right at home among the tropical foliage. Bright Lights chard has spectacular foliage and multicolored stems, in yellow, orange, red, and magenta. Unfortunately, we discovered that it can't take the heat, though it is a great edible ornamental for shadier or cooler locations.

We have grown some ornamental coleus just for the foliage but prefer the related oregano brujo. This species is a popular culinary herb in Puerto Rico, and some yellow-leaf forms are available. In recent years we have added some beautiful bulbs, notably gladiolus, to the front display. Just make sure not to plant daffodils, which are toxic, next to garlic chives, which they resemble. Once I went out to browse and saw that someone had harvested some daffodil greens, perhaps thinking they were garlic chives. I rushed in to warn Jonathan; thankfully, he'd figured it out on his own.

Our favorite groundcover in the front strip is sweet potatoes. We have grown many varieties, from ornamental yellow- and black-leaved types to edible clones like Beauregard. Most form enormous tubers, some football-sized. Some of the ornamentals like Marguerite make edible tubers, though they are a touch bitter to my taste. But Marguerite is a knockout when paired with red hibiscus and other colored foliage. Cooked sweet potato leaves are an important food in certain parts of the world. I was hoping that one of the ornamental clones would turn out to have delicious leaves, but the few I've tried have all been pretty bad. Selecting a more edible ornamental sweet potato would be a great contribution to gardening for some enterprising backyard breeder.

We have also raised some fantastic annual vines in the front yard, which really contribute to the tropical look and can produce a lot of food as well. Lablab, or hyacinth beans, are an outstanding ornamental with edible beans, pods, and flowers. We grow a purple-leaf form with pink flowers that has self-sown for years. One year we trained it up the guy wires for the telephone pole, and it grew about twenty-five feet high. When it came close to the top of the pole, the utility company had to come and cut it down because it was going to shut down power to the whole block. Since then we have managed our vigorous vines more carefully.

We have also grown bottle gourds to great dramatic effect. We train them over the sidewalk to form a living arch, and we have overheard several local teenagers telling their friends how much they like it as they walk by. Bottle gourds are remarkably vigorous and need constant pruning to keep them from swamping neighboring plants. But fortunately, the young leaves and shoot tips are a savory vegetable, so pruning labor doubles as harvesting.

From the very beginning, the tropical garden has brought us, our neighbors, and passersby a lot of enjoyment. It is remarkable what a little sheet mulch and a protected microclimate can do. Seeing those bananas, lablab beans, and sweet potatoes growing like gangbusters up there the first year gave us some hope that the rest of the garden would someday be a productive paradise as well.

By the end of the first year, our driveway looked like Puerto Rico. The hardy bananas grew to eleven feet tall and almost caused several traffic accidents as people slowed down to look at them as they drove by. (We started using them as a landmark for people coming to visit us for the first time.) And summer storms on the front porch were so much fun with banana leaves waving in the wind. When they were killed back by frost, we cut them down with a machete and piled on a few bags of mulch. The next spring we waited anxiously until they began to sprout in mid-May. In all of our eight seasons here, we only ever lost one over the winter. Since we have two, and each sends up lots of suckers, it was no problem to replant.

10

THE EDIBLE WATER GARDEN

By the time we moved to Holyoke, we already had three years of experience with edible water gardens, so in May of our first year, we were able to get right to work on a water garden using a plastic kiddie pool we'd had at Wonder Bread. We laid it down right on top of the fill, moved in potted aquatic plants that we had stored in the basement for the winter, and stocked it with mosquito-eating fish. Water gardens of this kind are a great strategy for compacted or contaminated soils, as they have no contact with the infertile or toxic layer below.

We'd found that the easiest way to grow vegetables in an edible water garden is in submerged pots. We tend to use pots that are low and wide—the opposite of most nursery pots—because many aquatic plants do not need deep soil but do like to run horizontally through the mud. The best soil mix is a blend of compost and sand, which is heavy enough to keep plants from blowing over but has good aeration and fertility. Although we mostly garden organically, it is hard to find organic fertilizers that will not dissolve in water. Dissolved water-soluble fertilizer would have fed algae instead of our plants, so we instead used nonorganic aquatic fertilizer tablets that we stuck into the pots with our fingers.

The centerpiece of our water garden, even while we were at Wonder Bread, has always been a Chinese water lotus. Like many ornamentals grown in water gardens in the United States, water lotus is an important vegetable in its homeland, but I have never met anyone here who grows it for food (though if you have ever had lotus roots in soup at a Chinese restaurant, it was from this species). The huge leaves are coated with a wax, and during a rainstorm or if you splash water on them, the water forms huge, beautiful droplets. The flowers are up to eight inches across, and after the petals have fallen, a seed head that looks like a shower nozzle forms. I had seen these in dried flower arrangements since I was a kid, but I never knew that this was where they came from. These nozzle heads are full of delicious nuts. If there is a finer ornamental food plant in the world, I don't know what it is.

Lotus roots are killed by freezing. Many books state that they are not hardy in Massachusetts. So imagine my surprise one day when, through a nursery day job I had, I was part of the renovation of a one-hundred-year-old lotus pond just down the road from me in Springfield. No one had told these plants that they could not grow in Massachusetts! As long as the roots are deep enough in the mud and are below the line where water freezes in winter, the plants can survive.

Another cold-hardy aquatic species that we eat a lot is water celery, or water dropwort. Like lotus, water celery is grown here in the United States as an ornamental but as a vegetable in Asia. In another strange parallel to lotus, water celery is reported to be hardy only in zone 9. I read one account on the Internet of someone's growing it in zone 6, so I thought I would give it a try. To our amazement, not only did it overwinter successfully but it even sent up shoots under the snow. By digging through the snow and brushing off the tiny two-inch shoots, we could have fresh microsalads in February in Massachusetts. The young leaves and stems of water celery are the main vegetable, and while we also grow this in soil on land, the leaves remain tender and good eating for a much longer part of the season when the plant is grown in water. We are glad we restrict it to an artificial pond, because it seems like something that could take over if it escaped into natural waterways.

We have also grown many other hardy water plants. Arrowhead has flowers and turniplike tubers. When you harvest them in the wild, you have to reach down into the mud with your toes to loosen up the tubers—in October. That's too chilly for me! But when you grow it in a pot, you just pull the pot out of the pond, turn it upside down, remove the pot, and there are all of your tubers ready to pick, just like turning out a pineapple upside-down cake.

Though we don't eat them that much, we also grow cattails, wild rice, and several other species. Some species we grow just to keep the algae under control. Native water lilies, for example, help shade the pond and outcompete the algae. Native submerged aquatic plants also would help take up nutrients that would otherwise go to algae (an unwanted weed in our system that can make the leaves of our aquatic crops gross and slimy).

In the early years of our Wonder Bread water garden, we grew lots of watercress, one of my favorite vegetables. But watercress requires flowing water to survive, so when we stopped using pumps to circulate the water, it refused to grow. We haven't quite said good-bye, though: we still grow some in our winter greenhouse each year.

Of the tender aquatic tropicals we've had experience with, water spinach is the best. The flavor of the leaves and young shoots, even raw, is absolutely to die for. It is easily grown and incredibly productive. I had read about it in *Edible Leaves of the Tropics* and purchased some at an Asian market. I rooted it in a glass of water just as I do for watercress and planted it out in our kiddie pool at Wonder Bread. Only later did I learn that this species is considered a noxious weed and that you are supposed to have a permit to grow it. You often see it in urban community gardens in Southeast Asian neighborhoods. Some immigrant farmers in Massachusetts have obtained permits and are growing water spinach commercially. Given that this species is killed by frost and does not even have time to flower let alone set seed here in our short growing season, banning its cultivation does not make a lot of sense to me.

The strangest and most remarked upon crop in our garden is water mimosa. This is another species from Southeast Asia. I have often wondered why so many aquatic vegetables come from Asia. Surely they can't have that many more tasty species growing wild in their wetlands than the rest of the world. I can only conclude that it has to do with rice paddies. Once you are already growing in a flooded field, it makes sense that you would begin to domesticate additional crops suited to those wet conditions. Water mimosa is a floating leaf crop that does not need to have its roots in soil to be able to grow. It comes with its own flotation devices, which look like someone was roasting marshmallows and put about ten of them in a row on a stick. Between each marshmallow is a bright red cluster of roots, where active colonies of nitrogen-fixing bacteria provide the fertility that makes the plant able to grow without soil. But that's not all. Touching the fernlike leaves causes them to close quickly, like sensitive plant. We have not been able to get water mimosa to overwinter, and it is expensive to order in the spring, but nothing else shows people the wonder of crop diversity like this floating, cabbage-flavored miracle.

When I fantasize about living somewhere with a longer growing season, raising my own water chestnuts is a major driver. Fresh-harvested water chestnuts are as superior to the canned product as sweet corn right off the stalk or a perfect vine-ripened tomato. They grow nicely for us in our pots-in-a-pond system, but they don't have quite a long enough season to ripen well here. The half-size tubers we have harvested were delicious but only served to remind me of what could have been. It would probably take starting the plants four to eight weeks early in a greenhouse before planting them out for us to get a good harvest. And somehow that time of year, we usually seem to have plenty of other things to do. We have also grown taro for the edible leaves and tubers in our water garden, but like water chestnuts, taro just really wants a longer season than we have to offer.

Producing food in our water garden is great, but to be honest that's almost the least of what we appreciate about it. Watching the fish, snails, and frogs going about their business there and dragonflies and songbirds coming for a drink, you can really see that a pond is the anchor of life in the garden. I'm sure that at night raccoons and possums and other animals visit it. Probably that woodchuck even comes by. That first year at Wonder Bread, we came out one summer morning to find a tiny tree frog sitting on our lotus leaves. The agroecosystem researcher in me was excited to see an insect-eating amphibian had been attracted by our provision of aquatic habitat. But the boy who grew up enchanted by nature is the one who dominated my feelings that day: I was so happy to see this tiny green life that we had made a home for.

One day I came home to our Holyoke garden and found that the pool was empty and the fish dead. Jonathan and I examined the kiddie pool, looking for a leak. Upon searching closely, we found a brick and some rocks at the bottom under all the foliage. We suspected our next-door neighbor's kids, who were the rock-throwing type. (I had been a rock-throwing kid myself.) They had wanted to see how big a splash they could make. Gardening in the city was going to present some new challenges.

Our neighbor brought her children over to apologize, and I think it was more uncomfortable for us than it was for them. We began a campaign to capture the hearts and minds of the children in our neighborhood. Our first step? We invited the rock-throwing family to come over for a tour of our garden. They ate strawberries and checked out our worms, and we told them our pond was about trying to grow food and provide habitat for fish and cool aquatic organisms, not just feeding us but helping our whole garden reduce pests and keep mosquitoes under control. Strawberries have remained a popular favorite with the neighborhood children. For years, one young man would ask us if the strawberries were ready—even in November. We sent a lot of those kids home with extra raspberry and strawberry plants for their own yards.

The experience reminded us both that good relationships with our neighbors are important and that social context is a key part of permaculture. Our goal was never just to have a nice garden for ourselves. We wanted to show our neighborhood and the world what permaculture can do, what an edible landscape looks and tastes like, and that easy and inexpensive techniques can heal terrible soils.

We thought about other ways to foster relationships with our neighbors, or at least not strain them. We tried to make sure that the pond wasn't a source of mosquitoes. The first year we ordered expensive, mosquito-eating gambusia fish. Then my friend Gerry, a coral propagator who runs an urban coral farm called Marine Reef Habitat in a greenhouse downtown, told us that goldfish work just as well. Now, every spring, I go to the pet store and buy "feeder" goldfish that cost about ten cents apiece and start eating mosquito larvae from the day they are put in the pond. We never feed them at all, but they grow fat and even reproduce on a diet of pond invertebrates and mosquito larvae. Our ponds have always been too small for fish to be able to survive the big freeze of winter. In good years we rescue most of them with a little net and give them away. Gerry would often take our fish and find them new homes with his customers.

During our second year in Holyoke, Jonathan and I decided to dig a proper pond. We wanted one that would be deep enough that a lotus could overwinter there, outside, so we would never again have to carry a three-foot-diameter pot full of heavy mud down to the basement to overwinter our lotus tubers.

At one of our work parties, we sketched out the area where we wanted the pond to be, got a few iron bars and pickaxes and shovels, and set a team to digging. By the end of the afternoon, we had completed excavation on our pond. Because pond plants want to be at different depths depending on their preferred habitat in the wild, we created a shallow terrace around the edge of the pond where plants preferring about eight to ten inches of water would be happy. Below that is a step about one and a half feet deep, and below that is the lotus pit a full three feet underwater, safely below the freeze line.

I went to Home Depot and bought a rubber pond liner. Jonathan scavenged some old rugs, which we laid down over the soil to prevent the rubber from being punctured by any rocks or sharp spots. Then the rubber liner went in, and we weighted it down around the edge with more salvaged stone. Around six p.m. we started to fill the pond with the hose. We had all been working since early in the morning, putting up our new trellis, digging a bamboo rhizome barrier, and making the pond happen. It was dark by the time it was full, but all of us were happy to get in and splash around.

11

URBAN FARMING IS MY DAY JOB

In 2003, right before Jonathan and I moved to Holyoke, Nuestras Raíces had completed construction of the Centro Agricola (Agriculture Center) project, a long renovation to turn an empty building and lot into our urban agriculture headquarters. The outside was painted with a beautiful mural and featured a plaza with edible landscaping including pawpaws and a water garden. Inside we built office and meeting spaces, a greenhouse, a commercial kitchen for community residents who wanted to start food businesses, and a restaurant. The project took three years and lots of donated materials. When we finished, it was time to breathe a minute and think what came next.

We engaged in a strategic planning process to decide what our big projects would be for the coming three to five years. The result was a mandate to start a farm. By this time members of Nuestras Raíces had ten gardens throughout the city. Many of our gardeners were already selling commercially on a small scale and were eager to expand. When a four-acre parcel of river bottom land two miles from downtown became available, we went for it.

When Nuestras Raíces director Daniel Ross told me that the person we'd hired to get the farm project up and running had quit, I decided to apply for the position myself. My Spanish had come a long way, thanks to nine years of board meetings. In my job at the New England Small Farm Institute, I had learned about the business side of farming and helped develop a course for aspiring farmers called "Exploring the Small Farm Dream," which was intended to crush their lifestyle fantasies and introduce the realities of farming as a business. I loved

NESFI but was ready to have a job with health insurance. Though I knew Nuestras Raíces would be hard work, I couldn't resist the challenge to put into practice all I had been learning at NESFI. Working close to home held a big appeal, too. I braced myself for a more-than-full-time job on top of the garden and completing two books.

On my first day of work in the autumn of 2004, I drove out by myself to look at the farm site. It was only a mile from my house but quite different. The farm sits on the banks of the Connecticut River and has wonderful deep soils with no rocks. At the time, it was overgrown, a tangle of half sumac and half goldenrod and bittersweet. I had never taken on a project of this magnitude before, and I was a little uncertain how we would get it clear, but I had a vision for what it would look like by the next spring.

Fortunately, I would be working with a community of people who had spent their youth cutting sugar cane. That first year we had five or six farming families signed up. We held a few machete festivals and cleared the site by hand. We dug out the roots with a marvelous bulldozer implement called a root rake, and suddenly we had an open field. We marked out six parcels, all either a quarter or half an acre, and threw down a fall cover crop of rye.

Winter didn't involve much slowing down because I had to have farmers ready to plant in spring. I threw together a Spanish-language farm-business-planning curriculum—no easy feat in a limited-literacy community with little business management experience. The following March we put up a little greenhouse in the snow, and by April we had broken out the rototiller and were getting down to business. I'm often asked why the Nuestras Raíces Farm focuses on "ordinary" organic annuals when I'm such a perennials guy. My job was not to do what I thought was cool but to meet the farmers where they were and implement their vision. Given that most of them had worked on chemical-intensive farms as laborers, getting to organic was a big enough step. We did some pretty fun edible landscaping around the edges of the farm, gardens, and downtown Centro Agricola that gave me room to play around a bit with my personal interest.

It was during the evenings and weekends, after coming home from long days of prepping the farm, that I put the finishing touches on our own garden design and got *Edible Forest Gardens* ready for publication. As soon as Dave and I submitted the manuscript in late 2004, I began completing and revising *Perennial Vegetables*. Meanwhile Jonathan was carrying much of the actual labor load in our garden. It's no wonder we didn't meet anyone to date right away.

MAGIC IN THE GARDEN
by Jonathan Bates

When we walk into the woods, many of us go to experience the solitude. We may want to be with the trees, birds, and other wildlife. We smell the freshness of the forest, damp and clean. We might notice fallen leaves, tree trunks crossing the path. Sometimes the earth feels spongy; it tends to give a little under our feet. I love to walk in the woods, and for many years I forgot about an integral part of that experience.

As a young person I used to go "hunting" at summer camp. Not for four-legged animals but for something much more elusive: mushrooms. Yes, I was lucky enough to go adventuring for old man in the woods, chicken mushroom, fairy ring, king bolete, destroying angel, and many others.

I loved growing and eating carrots one year. We dug them until the ground froze and then as soon as the snow melted away late winter dug some more. But did you know a close relative to carrot, poison hemlock, can kill you? Mushrooms, like plants, are not something to fear. Most are useful, even edible, and some are poisonous. We need to relearn where mushrooms fit into our lives.

I rekindled my relationship with the fungal world by going on wild mushroom forays. I bought some books on identifying mushrooms, found a local mushrooming club, and gleaned lore from experienced mushroom geeks. We hunted for and found all my childhood favorites and more. I learned to identify the poisonous ones, like destroying angel, first. My excitement was contagious. My dad, already a passionate bird watcher and food connoisseur, became a mushroom forager, too. Bringing family into the mushroom world with me doubled the fun. Then I created my deliciously edible and impossible-to-confuse-with-poisonous-lookalikes top five: king bolete, chanterelle, hen of the woods, chicken mushroom, and morel. Many food markets now sell these top five. Go buy some, see what they look and taste like, and invite mushroom geeks over for a meal and pick their brains. There are many good books on mushrooming, and I strongly suggest you read them before you start hunting, growing, cooking, and eating mushrooms. Soon you will be harvesting them yourself, and will never need to buy them again from the store.

I have transplanted some edible fungi into our vegetable patches. During the warm late summer months, a few days after a good rain, it is easy to drive, bike, or walk past a landscape and see mushrooms popping up all around. Many years ago I saw wine cap mushrooms coming up in a neighbor's mulch. I harvested a clump of mycelium-impregnated wood chips and brought them home to the garden. Within months and for years after we had our own wine caps to harvest.

We grow other mushrooms (shiitake, oyster) from purchased mushroom inoculant. Growing shiitake logs is an inexpensive and fun project, a surprisingly easy way to produce food in full shade. The solid wood logs marking the paths in our garden bear healthy, nutritious mushrooms for years, creating habitat for small animals and insects before eventually rotting down to build healthy soil.

Without mushrooms, compost wouldn't happen, mulch would not break down, and even some of our plants wouldn't amount to much. There is a thriving, exploding world beneath the soil, billions upon billions of life forms living and dying every minute. An entire kingdom, the fungi, are busy doing work for us all the time. Some live symbiotically with plant roots, sharing water and minerals in exchange for sugars. Some excrete enzymes to break down cellulose and lignin in dead wood.

Next time you venture into the woods or into your garden, remember the mushrooms. As mycelial messenger Paul Stamets states in the podcast *How Mushrooms Can Help Save the World,* "As [fungi] go across a habitat, they build food webs that support all sorts of other organisms that ride upon them. . . . These fungi are extremely powerful environmental healers. And when we engage them purposely, then they can be fantastic allies for helping us repair the ecosystems that we have so severely damaged. . . . These are the great soil magicians of nature." Let's bring more magic into our gardens.

12

THE GREENHOUSE: GETTING SERIOUS

During our second winter at Wonder Bread, Jonathan and I had purchased a kit greenhouse from Gardener's Supply Company. Twelve feet wide by sixteen feet long, it was a gardener's greenhouse, not a farm production structure. Neither Jonathan nor I had ever built a greenhouse, and it was our first serious construction job together. We felt that a simple project would be a good start for us, and in fact it let us chalk up an easy construction success. As much as we wanted an insulated greenhouse like Steve's at Tripple Brook, you don't build one of those on rented land you won't be staying on for long. Nor could we afford one. That dream would have to wait.

When we moved, we disassembled the greenhouse and brought it with us to Holyoke, where it sat in a pile in the yard until November 2004. The night after Jonathan and I finished reassembling it, I slept inside it, like a kid with a new treehouse, until it got pretty cold around three a.m. and I went back inside the house.

That winter we grew every kind of vegetable Eliot Coleman recommends in *The Winter Harvest Handbook*, from carrots and spinach to minutina and scallions. We planted late because of our tardy construction schedule, so we did not have good yields until late February 2005. But that's the challenge with winter vegetable growing in unheated greenhouses: you have to start your crops early, when it is still warm outside, which in Massachusetts means you have to seed them out by the end of September. During our first year, we didn't yet have a greenhouse in September; in the years that followed, we were still growing summer greenhouse crops in September. It is hard to pull out large, producing tomato plants from the greenhouse in order to seed mustards and spinach. Eventually we would start our winter crops in flats outside and transplant them into the greenhouse when the summer crops were ready to come out.

So what are our favorite winter greenhouse crops? All of the leaf crops come out tasting so sweet; they are of the highest quality when grown in a winter greenhouse. Spinach, arugula, sorrel, watercress, and Asian brassicas like mizuna and vitamin greens also grow wonderfully. Scallions, carrots, and beets hit a peak of flavor. And there's one crop that we don't have to plant at all: miner's lettuce, a sweet and tender green native to the western United States. This species enchanted us during our first winter in Holyoke, and we let a lot of it go to seed. Big mistake. Miner's lettuce is adapted to germinate in early winter and go to seed in early spring. It came up among everything else we planted in the greenhouse.

Winter in Massachusetts is depressing. We typically have snow on the ground from November through March, often one to two feet deep. While the greenhouse we built would never be large enough to produce huge salads for us to eat everyday all winter, it made a huge difference to cut a little fresh parsley or arugula into our soup and enjoy a plate of spinach and sweet watercress once a week through those long, dark days.

Our first winter in Holyoke, Jonathan and I sat down with our favorite edible landscaping catalogs and ordered hundreds of dollars' worth of fruits and berries. It was a moment I had been waiting for for a decade—the time to plant fruit trees I was going to stick around long enough to harvest from. In the spring we received one box after another filled with golden raspberry, grape, hazelnut, and persimmon plants. They arrived in a great diversity of sizes and forms, from six-inch-high tubelings to bare-root trees almost as tall as we were.

We started planting all the fruits we could fit, from low-maintenance natives like pawpaw, chinquapin bush chestnut, beach plum, and American persimmon, to tough introduced species like kiwi, hybrid hazel, and dwarf mulberry, to dainties including grapes, bush cherries, and mini-dwarf peach and apple trees. We also planted many kinds of berries, from golden raspberries to honeyberries, brambles, and blueberries. We created our private outdoor room with a mimosa tree, bamboo, and giant fuki.

Our first persimmon was so feeble it did not survive the first year. Jonathan and I were devastated. Thankfully, the nursery sent us a free replacement. We were also heartbroken when a roving groundhog chewed our dwarf mulberry down to the ground, though it restarted valiantly. Every setback meant another year until we tasted sweet, sweet fruit.

13

PERENNIAL VEGETABLE SPRING

After the long desolation of a Massachusetts winter, things start to move quickly once the snow thaws. Spring is the season of perennial vegetables, the time when their advantages over annual crops become clear. In the time it takes for annual crops to be ready to eat, many perennials have had three months of harvest. At that point, having bolted and lost their flavor until the coming fall or spring, the perennials pass the baton to the annuals.

As the snow melts, the bedraggled remains of the previous year's perennial vegetables poke out of the soil. Though there is some frost damage on the leaf tips, baby perennial greens will already be coming up here and there throughout the garden and will be ready to harvest by the last week in March.

One such green is sorrel, an often-overlooked vegetable. Partial shade, drought, and even two feet of snow seem to slow it down only temporarily. Sorrel is a great example of a multifunctional permaculture species because its deep roots concentrate nutrients from the subsoil. Calcium, phosphorus, and potassium accumulate in the leaves and are made available to neighboring plants over time as the leaves break down into the soil again. Sometimes Jonathan and I speed up this process up by cutting back our sorrels and mulching with them, causing a flush of tender new growth. Sorrel's sour leaves are good eating in spring but can quickly be lost to bolting and bitterness, although the nonflowering variety called Profusion keeps on cranking out fresh leaves all season long. Profusion sorrel is also notable for the density of its growth. We planted a row of it as a barrier between two different types of groundcover. Sorrel also grows nicely in the greenhouse all winter. Sometimes we transplant it in, but we had one in our hoophouse for many years with great success.

Like sorrel, water celery pokes its head up early with tender shoots as the snow melts. This only goes for the water celery we have planted in the ground; for some reason the ones in pots in the pond take much longer. From mid-March until the end of April, water celery greens are among our favorite types of greens for salads. Sometimes its parsley-celery flavor gets to be too strong by the end of May, but we're happy to have it in March and again in late fall. Jonathan and I get down on our hands and knees and graze on it like sheep. Our potted water celery plants can be weedy, but our patch in dry partial shade is well controlled, as it is out of its favored conditions there, slowing it down to the point that it often dies back in midsummer.

Early spring is also the time for Caucasian spinach shoots. Like a skinny asparagus with tender leaves, this is a high-class vegetable. We had trouble finding the perfect location and conditions for it, but one plant in somewhat moist partial shade has persisted for several years. By the end of April, Caucasian spinach shoot season has passed, but the leaves can be eaten raw well into June. Few perennial vegetables can compete with that lengthy season.

March is not too late to harvest last year's perennial root crops. Some, like sunchokes, are at their best this time of year. Also known as Jerusalem artichokes, sunchokes store their energy as starch in their tubers over the winter, but as spring comes they convert the starch to sugar in anticipation of the growing season. A decent vegetable in the fall, sunchokes become sweet as apples in the spring.

In early spring we dig other root crops, like skirret, Chinese artichoke, and groundnuts. It took me a while to become a fan of groundnut tubers, which seemed a second-rate substitute for potatoes. Although we grew plenty of them because they are native, high in protein, and fix nitrogen, it wasn't until I read Samuel Thayer's *The Forager's Harvest* that I learned how to deal with groundnuts in the kitchen. Thayer suggests treating them like the bean relative they are.

Following his advice, I mashed up some boiled groundnuts with chili spices and cheese as though they were refried beans. Suddenly this high-protein native crop found its place in my diet.

When April arrives, plants are leafing out everywhere in the garden. The delicacy of spring for us is ramps. This native wild leek grows in the shade of moist, deciduous woods throughout eastern North America. I've read about Appalachian ramp festivals where the whole town reeks of garlic for days. When we lived at Wonder Bread, Jonathan and I decided to host our first ramp festival, a tradition we continued four or five years. This celebration of spring abundance drew friends with dishes like nettle quiche and Japanese knotweed crisp; our friend Frank Hsieh once brought a whole roasted spring lamb from his farm.

I had been keeping an eye on a patch of ramps across the street from a Subway sandwich shop the next town over for several years, and my sustainable harvests (never more than 5 percent of plants, as ramps grow slowly) had provided the ramps for many of our festivals. One year I drove by to visit the ramps and saw, to my horror, that a large condominium development was going in. I rounded up a crew from the ramp festival and we rescued hundreds of plants. Today they and their progeny are growing in our own garden and many other gardens as well.

Perennial scallions come into their own in April, too. If you grow scallions from seed, you can plan the harvest in order to have them any time of year. That's nice but more work than Jonathan and I had in mind. Every spring our Welsh and walking onions send up their new scallions. We dig or divide their clumps for harvest and transplanting. This glut of scallions is welcome after a long winter, and we put it to use in fried rice and scrambled eggs. Fall brings a second flush of scallions.

At the base of our bamboo, in the shade of a feathery-leafed mimosa, grows a plant with enormous round leaves up to three feet across. This is fuki, a popular wild edible in Japan and a bold statement in the landscape. Not content to grow ordinary fuki with its eighteen-inch leaves, we obtained the giant form from an alpine plant nursery, of all places. Giant fuki is sterile and thus has no chance of dispersing into the environment—except for its aggressive rhizomes. Our fuki is hemmed in by a bamboo rhizome barrier on two sides and a frequently trodden path on the other. We actually wish it would grow faster so that we could harvest more, because fuki is a favorite spring vegetable on both sides of our duplex. You eat the leaf stalk, as you do rhubarb, but fuki is more analogous to celery. We boil the stalks, peel them by hand, and marinate them in umeboshi or raspberry vinegar with some shredded ginger and tamari.

Some people balk at the labor involved in preparing a crop like fuki. I don't mind spending fifteen extra minutes processing a vegetable that, as a perennial, takes no work to grow beyond its first year. It's a question of which stage in the growing process you want to labor for your food.

April brings an embarrassment of perennial vegetable riches that just keep coming. Asparagus, good King Henry, and giant Solomon's seal provide tasty shoots. Even the tightly curled shoots of hostas can be cooked up this time of year; they're not my favorite, but you'd be hard-pressed to find a hardier vegetable for full shade. Violet leaves are usually fairly bland, but Dave Jacke introduced me to a cultivar called Rebecca that has remarkable vanilla-mint-flavored leaves and cream-and-purple flowers. Another April favorite is the wasabi-like Eastern native toothwort, a groundcover with piquant horseradish-flavored roots and leaves. I think a commercial producer could market the roots to the finest sushi restaurants.

The king of cold-hardy perennial vegetables is asparagus, and around here May is the peak of harvest. We grow a variety called Purple Passion, and its fat spears keep coming for more than a month. Fresh-steamed asparagus is as good an argument for turning your lawn into a garden as any I can imagine.

An unlikely favorite of ours is sweet cicely, which we fell in love with at Wonder Bread after discovering that when the large seeds are green and unripe they taste like licorice. We all eat a lot of these seeds in the spring, and they also became a favorite among all the children who visited the garden. Jonathan and I had tried to start sweet cicely from seed when we lived at Wonder Bread but had never got it to germinate. We finally bought some plants at Wonder Bread, transplanted them to Holyoke, and when they set seeds for the first time threw fresh seed all over the garden hoping that a few might germinate. It seemed as though every single one did, creating a weed problem for us. We began to deadhead what we didn't eat to keep it under control.

During those first couple of years in Holyoke, Jonathan and I were both still single, and we spent an inordinate amount of time in winter (and in the summer, after dark) reading up on useful plants. Many a time one of us would cry, "Dude!" and run over to learn the details of some strange crop on the Plants for a Future database or from a moldy old tome. One was a perennial arugula called sylvetta I had profiled in *Perennial Vegetables*. I filed it away in my head as hardy only to zone 7 and dismissed it as a candidate for our garden. But Jonathan insisted he wanted to try it, and I went along, thinking to myself that it didn't stand a chance. It grew vigorously as an annual, and to my surprise the plants resprouted the next spring: The woody parts of this shrub are not hardy, but the roots survived, and the leaves are the part you eat anyway. (So much for my award-winning expertise.) Sylvetta went on to self-sow with abandon until we corralled it under our grapes, where it can do its thing without smothering anything more delicate. The strong arugula flavor of sylvetta is outstanding in omelets, but perhaps it is at its best chewed fresh in the garden with a few ripe alpine strawberries.

The perennial vegetable with the longest season in our garden is garlic chives. Most people in the United States view this humble plant as a minor crop at best, or even exclusively as an ornamental. But in China the blanched shoots and flower stalks are a commonplace crop. Jonathan and I love the garlicky greens in the spring and fall, even though we have many other perennial vegetables to choose from at that time. But in late summer this crop comes into its own with twelve-inch stalks topped by an edible flower bud. The full stalk and bud can be chopped and thrown into whatever you might be cooking for lunch or dinner. I often see bunches of these flower stalks for sale at Asian markets, but I rarely meet a gardener who uses their garlic chives in cooking. Once they open, the flowers are also quite attractive to honeybees. As with sweet cicely, we could not get it to grow from seed packets, but fresh-sown seed turned into a weedy disaster. We have now isolated our garlic chives in some areas that it can dominate (and be deadheaded) and ruthlessly weed it out of our other beds.

As we were establishing the garden, we relied on self-sowing annual and biennial plants, and they became important in our diets. These species aren't *exactly* "self"-sowing—we usually scatter the ripe seeds and fruits around the garden wherever there's an empty space or disturbed soil.

Our favorite self-sowing species is Western Front kale. This variety looks like Red Russian and shares Red Russian's sea-green color, oakleaf form, and mild, creamy taste. We ordered seeds of Western Front when we lived at Wonder Bread, hoping for a perennial kale (we had already successfully winterkilled Tree Collards kale and Dorbenton's perennial kale). Actually, the seed company we bought them from claimed that about half of the seeds would turn out to be perennial, as it was still being bred for this. We planted out about a hundred at Wonder Bread, but the perennial seedlings, if there were any, did not survive the winter. (This is not the only crop where perennial genes are linked with frost-tenderness.) However, in growing it we discovered that we loved this kale, and when it flowered and set seed its second year, we let it dry on the plant and scattered it throughout the garden. That fall hundreds of tiny, tender kales came up around the garden and started a tradition that we practice to this day. This kind of repeated seed saving and sowing is a form of passive plant breeding. Each year it is possible to save seed only from the plants that survived minimal enough care to reproduce, and so over time we have been developing a semiferal strain of kale that can find and exploit available niches in our mulched perennial garden system.

We have given away lots of seed over the years, and every now and then I see some of our kale's descendants thriving beneath an Asian pear or at the edge of a path in a friend's garden.

Yellow Pear cherry tomatoes look like miniature yellow light bulbs and are mild and sweet. Throwing some fruits around in fall effectively plants the following year's crop. We phased this out after a few years in Holyoke not because Yellow Pear wouldn't grow but because the plants got so large that they were crowding out our perennials. The other disadvantage is that when they grow from seed in spring (as opposed to using established transplants), the fruits don't ripen until late in the season and then have only a short window before frost. The same holds true for tomatillos and annual ground-cherries.

Sometimes I get a little cocky and start to think I know everything there is to know about useful plants for my region. So imagine my surprise when one day I stumbled across a reference to the edibility of the species I had always read was poisonous. Black nightshade is a native tomato relative with white flowers, broad green leaves, and tiny black fruits that look like microeggplants. I found a reference to the edibility of the fruit, and as it turns out Jonathan and I found some growing on the street-facing side of a neighbor's hedge. Suddenly I began to see it in empty lots all around me, coming up between the cracks in sidewalks. Here is an edible fruit that had been under my feet for years. The fruits are small but tasty, sort of like a licorice-flavored tomato. Apparently the unripe fruits are toxic, which perhaps accounts for its bad reputation. We brought home some fruit and sprinkled it around the garden, and the next spring a few plants came up. Once again I threw the little fruits all over, and we were soon inundated with far more black nightshade fruit than we could possibly use.

Some things we felt certain would self-sow utterly failed. Miner's lettuce, though a terrible weed in our greenhouse, never established itself outside. The same went for cilantro and dill, both of which we would have loved to have more of as culinary herbs and sources of nectar for beneficial insects. Meanwhile we were surprised to find successful seedlings of eggplant, lablab beans, purple mustard, mache, and even castor beans. It had never occurred to me that tropical species like lablab and castor would have seeds that could survive a Massachusetts winter.

Besides self-sowers, the annuals we grow alongside our perennial staples are those that are at their best fresh from the garden, like peppers and heirloom tomatoes and annual greens, plus species we use in favorite recipes, like garlic, basil for pesto, and tomatillos for roasted green salsa.

BROCCOLITAS FOREVER!
by Jonathan Bates

When I first saw sea kale, I fell in love. Against the dark green foliage of the rest of the garden, the powdery blue-green and purple hues of new leaves called out. The tender spring leaves taste a little like collard greens, but a single plant yields a crop of only a half dozen leaves a year; eating more would put an end to the plant in short order. One alternative is to eat some leaves in the fall after most of the growing has taken place, but we have come to prefer the early spring broccolis. Along with the first six to eight inches of tender new flower stalk growth, the broccolis, or broccolitas, can be eaten raw, mixed into salads, lightly cooked with butter and salt, or added to a vegetable stir-fry.

Perennial broccolis like sea kale can be grown as permanent, low-maintenance, early-season vegetables. Like asparagus, sea kale broccolis are ready to harvest weeks before most annual vegetables can even be planted, extending the growing season. Our sea kale has lasted for ten years, and we've been able to enjoy its flowering stage every spring. Imagine an explosion of three-foot-tall, snow-white bundles of small flowers filling the garden with a honey scent that attracts bees. Take that, annual broccoli!

Sea kale is not the only broccolita in our garden. Turkish rocket also forms a pungent, mustardy-tasting broccoli raab that is best cooked, lending it a nutty flavor. The plant has a deep taproot that helps support it through drought and mines the subsoil for important minerals. The mature hairy leaves protect it from most pests. It is a long-lived perennial, increasing its food mass as it ages. Turkish rocket's nutritional value is high; its 22 percent crude protein content at flower-bud stage is comparable to peanuts.

Both of these broccolitas are from wild lineages. Once established, these plants are fairly pest and disease resistant. (North American pests haven't figured out how to get through sea kale's thick, waxy leaves, for example.) And they seem well adapted to drought. The downside is that neither plant has really been domesticated yet; they need breeding work to produce bigger broccolis.

I often wonder why more people aren't growing perennial vegetables like sea kale and Turkish rocket. A monoculture of annual broccoli may grow more calories than a monoculture of asparagus or sea kale, but how much soil is eroding from the plowing and cultivation of annual vegetables? How many chemicals and how much fuel and water do they require? Perennial veggies challenge the monoculture mindset and the industrial, turn-a-profit-fast economies of the scale food system; they can be grown with less labor and fewer inputs and with zero soil erosion. When grown with a diversity of other perennial vegetables, fruits, and nut-producing trees and shrubs in multistory polycultures, the long-term benefits and yields of a plant like sea kale outweigh any perceived shortcoming in yield.

I challenge willing gardeners to tuck these exciting plants into perennial beds to gain a little more fresh spring eating. Enjoy the broccolitas!

14

PUTTING DOWN ROOTS AND A PARAKEET VISIT

T hough we experienced fast and dramatic results in our tropical and edible water gardens, in most of the rest of the garden, growth was slower and more sporadic. That first year (and for a few thereafter) the tallest plants in our backyard were annuals—something of an embarrassment for a guy who writes books about perennial edible forest gardens and set out to prove their potential. There simply wasn't much "forest." The few birds that visited the garden during that time perched on stalks of corn and amaranth; Jonathan and I imagined them looking down at our foot-tall fruit trees in disdain. Meanwhile our disassembled greenhouse had been sitting in a pile in a back corner of the garden.

From the beginning, Jonathan and I struggled against weeds. In part this reflected the empty spaces that still filled a lot of the ground between our plants, but the real problem was that weeds were coming in on our compost and straw mulch. It would be years before we found a weed-free compost source and straw mulch free of oat grass. Our very efforts to mulch and eliminate weeds were themselves bringing in more weeds!

One day both of the dwarf cherries we had planted right away began to wilt. They had plenty of water and nothing else in the garden was wilting, so Jonathan and I were at a loss to explain the problem. We scrutinized them and spotted a jellylike sap with sawdust coming out of holes at the base of the trunks. We knew what this was: the work of the peach tree borer, a devastating pest that kills stone fruit trees by tunneling completely around the trunk. Our little cherry trunks were only an inch in diameter, so we knew this was probably a fatal event. When we examined the other stone fruit trees we'd planted—Nanking cherries, beach plums, and peaches—we saw the same sap and sawdust.

We went after the grubs with wire to remove them and contain their damage. This was the first test of our approach to pest control. Would we be able to monitor and prevent these borers? Or would we remove our susceptible stone fruits and plant more resistant fruit trees, such as seaberry or goumi?

Some of both, as it turned out. We fought to keep our mini-dwarf patio peach until it was clear that squirrels, insects, and a multitude of diseases were going to make it unworkable for us. We were able to nurse one of our beach plums back to life, and the other resprouted. But we couldn't keep the cherries healthy, and with regret we took them out. Since then we have learned that in a healthy soil and with sufficient irrigation, stone fruits resist borers much more effectively. In fact, borers would cease to be a problem for us in a few years, once our soil was healthier.

Though we did not right away see the kind of habitats we'd envisioned (the garden was simply not yet a functioning ecosystem), by August of our first year there were many hopeful signs of the garden that was to come. In 2004, when corn was our tallest plant and before there were woody plants for birds to perch on, we had a visit from a most unlikely visitor: a parakeet. Though it had probably just escaped from someone's house, to Jonathan and me it seemed to be a messenger coming to tell us that an edible paradise was on the way. The parakeet came close to us as it poked around in the soil, looking for bugs and seeds for several hours before it took off, never to be seen in our garden again. Later that year we had a visit from a wild turkey. While turkeys are common in rural Massachusetts, to see one in an urban neighborhood is highly unusual. We felt that these visitors were coming to check out our little project, perhaps as emissaries of the larger ecosystem.

Jonathan and I set out with a rigid ecofunctional approach to the garden. Except for the hardy bananas in the front yard, we didn't plant ornamentals; everything had to have a practical function. So it was a pleasant surprise when the sweet cicely, skirret, aster, and other flowers we planted to attract beneficial insects turned out to be beautiful as well. We also hadn't considered that anything that makes a fruit makes a flower first. One spring morning I looked out the window and saw a multitude of herbs, young shrubs, and even our straggly little apple tree covered in flowers. Our sweet cicely, earth chestnut, and Scotch lovage, members of the umbel family, created cloud-like drifts of white, cream, and yellow; the coreopsis and green and golds of the aster family, which we had also planted for pest control, added hundreds of little daisylike flowers. The pealike flowers of the legumes we had planted for nitrogen fixation and the flowers of fruits and berries like pears and raspberries made a pleasing habitat for us as well as the pollinators, parasitoids, and predators we intended it for. Seeing all this beauty started to soften my pragmatic, ecofunctional, plant-geek heart. The garden was already changing me.

We began to see what the seasonal flow of garden was going to look like. In winter we would eat salad greens from the greenhouse and obsess over catalogs. Spring would be the season of perennial vegetables. In summer we would switch over to annual vegetables and berries. Fall would bring (some day) tree fruit, nuts, root crops, and fall vegetables.

15

MEANWHILE BACK AT THE FARM

By the end of 2005, I was learning to juggle the complexities of the Nuestras Raíces farm project. There, fertility and pests were the least of my worries. I had to deal with getting permits from the city, writing reports to funders, and dealing with complicated interpersonal dynamics among my farmers.

Next to our four-acre farm, there was a twenty-six-acre parcel of land. Like our farm, it abutted the river and had a stretch of fertile river bottom loam soil. It was owned by the nuns of the Sisters of Providence, and various local characters had told us that they would never let us farm it. Never one to be told something is impossible, Daniel went to talk to Sister Joan, the head of the order. She was interested, and we began a process of negotiating for access to the land. After many months of negotiation and paperwork, Nuestras Raíces came away with a year-to-year agreement that would allow us to start farming in spring 2007. Our rent? A dollar a year. Today the land is held by the Trustees of Reservations, and a long-term lease is under development.

This was going to change the farm project—radically. I had two years of learning to keep four acres together, and here we were dramatically expanding acreage. But you don't turn down the chance to increase your acreage by 500 percent for a dollar. We would be able to have much larger parcels for our farmers and greatly increase the number of farmers as well.

Fresh off of completing the design for our home garden, I decided to take on the challenge not only of another and bigger farm but a participatory design project for the expanded farm. I coordinated a group process, the first phase of which was to develop a list of goals. We convened some of our farmer-participants, youth, board members and other stakeholders and generated a list of things we wanted to see happen on the farm. This included larger commercial agriculture plots and associated infrastructure such as a well, barns, fencing, and greenhouses. We also wanted the farm to come alive as a cultural agritourism destination. A couple friends of mine, Keith Zaltzberg and Bas Gutwein, needed final projects for their landscape design degrees at the University of Massachusetts-Amherst; they turned the results of our process into maps.

Our team performed a site analysis of the farm, just as Jonathan and I had done at home. In this case, and at this scale, the key issues were not sun and soil but rather floodplain and legal restrictions. We created a map that showed the different zones and the constraints on their use. Keith and Bas came up with cardboard cutouts to scale representing the elements included in our goals: a forty-by-forty-foot barn and multiple one-acre farm parcels, greenhouses, and all the other components we had listed.

We convened a series of smaller groups: one for the farmers, one for youth, and one for the board of Nuestras Raíces. Each group placed the cardboard cutouts to represent how they felt the site should be designed. We photographed each group's design, and Keith and Bas then made maps from each of them, which we presented to a final, combined large group. Participants voted for the design elements they liked best. Everyone was in agreement about the basic layout of Nuestras Raíces Farm. Of all the things I did during the time I worked for Nuestras Raíces, this participatory permaculture design is one that makes me proudest. It showed me that you don't need a fancy degree or even a permaculture design certificate in order to design.

A few things also happened at the farm that I'm not so proud of. Nonprofits are always looking for inexpensive or, better yet, free resources. Shortly after we agreed on the design strategy, we started looking for the resources to put it into place. My team and I took down and relocated to the new, combined thirty-acre farm three green-houses, several sheds, and a timber frame barn dating to the 1870s. We got permission to go into a condemned restaurant and pulled out tens of thousands of dollars' worth of kitchen equipment, including two working walk-in freezers. Nuestras Raíces could mobilize a crew quickly, and we had a dump truck and a tractor with a loader to take advantage of this kind of opportunity.

So when someone called and offered the farm a pair of free goats, of course we said yes. It turned out these goats had been locked in a shed for most of their lives and were unsocialized, so when they got to the farm they wasted no time in hopping the electric fence. Farm animals had got out of their enclosures before, in our four-acre days, but they were never a big deal. I learned, for example, that there's almost no way to catch an escaped pig until it tires, at which point you have to pick it up or drag it back. A few of our pampered goats had escaped but came back when we rattled feed around in their bowl. These new goats, though, didn't know who we were. They struck out across the highway that bordered the farm and headed up into a steep, dense, ten-acre patch of poison ivy, brambles, and sumac.

That night Daniel, Nuestras Raíces youth director William Aponte, and I bushwhacked up through the woods with flashlights trying to find the goats. This would have been one thing if we lived in a rural area. But our farm was half a mile from the largest mall in western New England. My biggest worry was that the goats would cross the highway and that someone would be hurt in a car accident.

For several weeks, the goats decimated flowerbeds from neighboring yards and frightened children at a nearby daycare center. Fortunately, they seemed to really like eating poison ivy, and we realized they were for the most part staying in their thicket. Several people helped us track them down. The Galarzas, one of our founding farm families, lay in wait and established that the goats came out of this forest to graze on the lawn of the daycare center at the edge of the ten-acre thicket every morning around six. Agnori and Pedro, two of my livestock farmers, tried and failed to lasso the goats on several occasions but then came up with an excellent idea for a trap. Using sturdy portable cattle panels (extremely versatile farm essentials, cattle or no), we built an enclosure sixteen feet wide by sixteen feet long, with extra cattle panels on top to completely enclose it. We put a portable plastic calf hutch inside for shelter if it was going to be a long wait. The gate was a pallet rigged up so that it would snap shut when we pulled on a thirty-foot string from a hiding place behind a nearby tree.

To complete our literal stakeout, we tethered a female goat inside, set out some goat feed, and settled in behind the tree to wait. Pretty soon one of our runaways came out of the woods and walked right into our trap. The other goat soon followed. We knew we couldn't bring two goats with such bad habits back to our farm. But they tasted excellent! That's about as free range as a goat can get.

Lining up educational opportunities in Spanish for my farmers was part of my job at Nuestras Raíces. Many of them wanted to learn about livestock housing and health; as a plant guy, I was in over my head. What to do?

We had a resource available about ninety minutes from Holyoke, in Rutland, Massachusetts. There, Heifer International runs the Learning Center at Overlook Farm, one of four campuses in the United States that promote its mission. Heifer is a development NGO that gives livestock and trees to farmers around the world with the provision that they pass on the gift of offspring or seedlings to another farmer within a few years. I had asked a Heifer staff member if they had any connections that might help Nuestras Raíces. Heifer invited my Spanish-speaking farmer-participants and me to visit the farm for a tour and also offered to provide translation.

One snowy day my farmers and I drove out to Rutland. As we sat in their meeting room waiting for our tour to begin, a woman with brown eyes and the most beautiful dark hair I had ever seen walked into the room. I spent the next half hour trying to see if she was wearing a wedding ring.

Until that moment, dating in my thirties had felt like a string of Seinfeld episodes. I had a ridiculously long list of criteria of what I was looking for in a woman, and it seemed impossible for anyone to meet them. Even the women I went on a few dates with never worked out; one dumped me for her dog, a recent rescue who didn't like men. Just a few weeks before I went to Heifer, with the help of my expert dating advisory team (friends and counselors who had actually found someone themselves), I narrowed my list down: someone warm and caring who speaks Spanish and likes gardening.

Marikler Girón Ramirez was our translator for the veterinary training that day, though I confess I didn't pay much attention to details of goat health. But I did resolve that I was going to ask this woman out, no matter how excruciatingly awkward it was. I consulted Milagros Guzman, a Nuestras Raíces youth leader; without her teenage love advice, I'd probably still be single.

I called up Marikler a few days later—as ordered. I was intent on asking her out in Spanish, even though my Spanish was still not great when it came to talking about things other than crop farming or board meetings. I was so nervous I couldn't follow everything she was saying. All I know is that I got a yes.

I was going to see Marikler the week after we first met anyway because we were due to translate together for a seed-saving workshop for my farmers and community gardeners from several urban organizations in the state, hosted by Heifer and featuring a UMass professor as the speaker. A day after asking her out, I found the excuse to make a thorough list of seed-saving terms and go over it several times on the phone with Marikler. We did manage to pay enough attention to the workshop to do a nice job of translation, despite all the talk about pistils and stamens and pollination. Our first proper date was the following weekend.

Marikler grew up outside of Guatemala City. Committed to improving conditions for the poorest people in her country and elsewhere, she took a job with Heifer International, doing everything from administrative work to organizing meetings for representatives from across the Americas. After eight years of working for Heifer in Guatemala, Marikler came to the United States to share stories of their work on the ground with funders here. She was asked to join Heifer's fundraising office as a full-time staffer in 2003 and had been there ever since. Fortunately for me, her duties included the occasional translation work.

I'm not above admitting that besides her obvious commitment to social justice, I was also impressed by the practical agricultural skills Marikler brought to our relationship. Not a plant geek like Jonathan and me, Marikler had experience with farm animals. She grew up with chickens, ducks, quail, and pigeons as meat animals in her urban backyard. She could slaughter and skin a rabbit and prepare the pelt, had assisted with the deliveries of goats and sheep, and had cared for yaks, llamas, and water buffalo. She could also diagnose intestinal parasites by looking at the inside of a goat or sheep's eyelid. My jaw dropped when she mentioned on our first date that she could do tricks while riding on the back of a draft horse. Here was somebody who not only met my list of criteria but blew it out of the water.

Marikler's warmth thawed my heart. I thought I had given up on love and was resigned to the idea of spending my remaining years with dusty old plant books. But being with Marikler made every part of my life better, and still does. As she and I were falling in love, Jonathan's former crush, Meg, came back into his life. Suddenly the release of a new, improved perennial grain variety was not the most important event in the house.

BUILD IT AND THEY WILL COME
by Jonathan Bates

Many forms of life on earth build structures from materials in their environment to protect themselves or to attract mates. A gorilla uses large leaves to cover himself from heavy rain. A male bowerbird builds an elaborate nest and lines it with colorful objects to attract a female. An aquatic caddis fly larva forms a tube of sand around its fleshy body to hide from predators. Even bacteria have evolved to live in the swollen parts of plant roots. Called rhizobia, they live symbiotically with the plant and fix nitrogen, a perfect match indeed. Humans have also evolved not only to build structures for ourselves but to create one-of-a-kind, shiny houses, like a bowerbird's nest, to attract mates and shelter our families.

My first foray into ecological building came about when Eric and I needed a toolshed. Rather than a simple and convenient big-box-store prefab metal shed, we considered a more complex and adventurous project: building a shed from scratch from locally harvested materials.

First I had to convince myself that I could build one. I knew how to use basic tools like a hammer and saw. But we wanted a shed that would last and look nice. Could I create such a structure? How much would it cost? Where would I find the materials?

My design was simple: a ten-by-ten shed with a double barn door made with 95 percent natural, local, or recycled materials. I acquired reused patio pavers for footers, rot-resistant black locust as the foundation, waste pine slabs for the east wall, reused plywood for the west wall, nicely milled one-by-six hemlock boards for the south wall and floor, and deconstructed oak pallet boards for the north side. The interior framing was a mix of two-by-fours of oak, hemlock, maple, and pine. All of the wood I purchased came from a local ecovillage microsawmill. I found another local microsawmill that specialized in milling black locust.

Growing black locust is banned in Massachusetts due to its "invasive" nature, but it was by far my best option for a local, rot-resistant, wood foundation, far superior to toxic pressure-treated wood. This abundant tree native to nearby Pennsylvania was screaming in my ear, "Here I am! There's lots of me around! I'm a climate-change fighter! I'm rot-resistant! I fix nitrogen and build soil, and you can eat my flowers, too!" How could I ignore such a fine natural building material?

The ecological nature of this project didn't stop with the building. For months after it was completed, a number of skunks visited our garden and home—sometimes living under the shed. Waddling through the landscape, grubbing for earthbound food, they were like little pigs, digging small holes all over the garden, sometimes killing rare and precious plants we had only just put in. This might seem harmless, but the skunks were not welcome. I watched them till our soil and mulch, opening up the ground for weeds to grow. My anger built up, the stress blinding me to the ecological realities.

How could we solve the skunk problem? Poison them, shoot them, trap them, or chase them? All of these options ignored the obvious. The skunks liked the bounty we had created, and nothing we did to get rid of one would stop new skunks from filling the void. Eventually I mellowed to their visits and asked myself different questions. How were they getting here? Where were they coming from? These simple questions, arising from curiosity rather than frustration, allowed me to come up with better solutions.

The solution was to put up short barriers at their entry and exit points to and from the yard and to keep them from nesting under the shed. Minor adjustments can create powerful solutions. It took me a few hours to build some simple fences at the back of the yard and to wrap chicken wire around the base of the shed. I see much less evidence of their presence than before.

So I learned I can build a shed and discourage skunks, but would my nest attract a mate?

I had not been to a bar for over a year; drinking alcohol lost its appeal for me in college. But I wanted to spend time with a friend, so we decided to see a show at a local bar neither of us had been to. It was a good show with great music. During the band's break I felt a tap on my shoulder. "Hey, is that Jonathan Bates?" I turned around. A long-lost friend had found me. We talked, but it wasn't until later that night that I remembered her name, Meg.

Meg had just returned from Mexico, where she had been teaching violin for a year. We had lost touch while she was away. (In reality I conveniently forgot about her after she "just wanted to stay friends" five years before.) But this new encounter rekindled the friendship. We were both dating other people at the time, but I found it curious when she asked me to lunch, and on those first few dates I realized both of us had changed. There was a sparkle in our eyes that hadn't been there before.

I didn't originally build the shed for the skunks to make a home under. I built it as a functional example of how I wanted to live my life, with ecological integrity, and lasting strength. Thank goodness I rid the garden of those skunks, because two years after she tapped me on the shoulder that night at the bar, Meg and I got married in the garden. Finally, I had found my mate, and the shed and the garden weren't needed as attractions after all.

16

===

A MOVEMENT GERMINATES

If Jonathan and I could start again, we would have done a few things differently. Most important, we would have done something about our compacted soil. Laying sheet mulch on top of compacted soil had about the same results as making a compost pile on top of concrete, which is to say you can grow some things nicely, but it is not ideal for trees and crops with deep roots. In fifty years worms would have loosened up the soil, but we didn't have that kind of time. We did look into hiring someone with a tractor and subsoil plow, but it seemed silly for such a small garden. We also didn't think that kind of equipment could fit through our alley.

We also lost an opportunity to easily fence out wildlife from the garden. That simple step, with a couple of gates on the alleys, could have kept out skunks and a few pestiferous groundhogs. After you have plants growing everywhere, strapping chicken wire tightly to the existing wrought-iron fence is tricky.

We also wish we had installed an irrigation system. We had been hoping to capture rainwater from the roof and store it in tanks, running the water downhill to the plants using gravity pressure only. But our site was exactly the wrong configuration for this. We received some free soaker hoses and set them up as a stopgap measure that lasted several years, until they started to leak. It would take ten to fifteen barrels to hold all the water we would need in a drought, and it is difficult to find a place to store them. Perhaps between stacking some under the front porch and tucking a few behind the shed we might yet pull it off. If we had just invested some time and money on an irrigation system early on, it would have been so much easier than working around all our existing plants and infrastructure today. But it was so much more fun to plant fruit trees!

Almost as soon as we bought our greenhouse, Jonathan and I were fantasizing about one day having a subtropical greenhouse. Although they are unheated, Tripple Brook's insulated greenhouses never get colder than 20°F, even when it is −20°F outside. That means you can grow all kinds of phenomenal subtropical and Mediterranean crops. Fresh strawberry guavas (a tasty, dwarf guava relative) and a steamy greenhouse in the backyard is an alternative to a tropical vacation. We would be able to raise tilapia instead of goldfish and aquaponic vegetables and food year-round. However, we knew during those first few years that we already had too much on our plate and lacked the money and knowledge to undertake such a project. We simply couldn't do everything.

So we contented ourselves with raising tropical crops in the greenhouse during the summer. An unheated greenhouse does not give you many more weeks of frost-free growing—perhaps two to three weeks on either end of the season. Still, fresh tomatoes in early November, almost a month after the last ones are ripe outside, would be enough to make it worth it. We've tried other crops in the summer greenhouse besides tomatoes, including peanuts (they did not yield well) and several obscure root crops from the Andes. Oca was just starting to form tiny tubers when it was killed by frost. Mashua, a relative of nasturtiums grown for its edible tubers, didn't like the heat of summer. But we did have luck with yacon, a sunflower relative that can produce huge, crisp tubers and for us yielded better than sweet potatoes. Another month frost-free and it would have been a remarkable crop indeed.

A final regret is an error that Jonathan and I made many times— throwing around seed of a species we wanted more of only to find that it became a weed itself. In one case, however, a bit of weeding was the least of our worries. Cow parsnip is a native umbel family species that has multiple edible uses. The absolutely enormous flower clusters are dramatic in spring and fantastic for attracting beneficial insects. I had read many times that cow parsnip can cause a sort of hypersensitivity to sunburn, but as Jonathan and I had never experienced such a thing, I discounted it—until one day when Jonathan was clearing the pathways and chopped back some cow parsnip foliage. It was a hot day, and he had no shirt on. He got sap on himself and developed painful boils that lasted for several weeks. We began a campaign of cow parsnip eradication, though we have created a reserve in a corner where it can grow without endangering people, as its many uses are still too tempting to get rid of it altogether.

Despite our regrets, it was clear by this time that we were doing something right and that life in general was moving in the right direction. Dave and I had been working on *Edible Forest Gardens* for almost nine years. It had become a perpetual project that my friends teased me about. I had come to cringe in advance of family obligations where I would need to answer questions about it, and at times I despaired that our work would ever see the light of day. But the book did at long last go into production.

In late summer of 2005, Dave invited me to speak at a permaculture design course he was teaching in nearby Shutesbury, Massachusetts. When I arrived, he opened a box and took out my very own copy of *Edible Forest Gardens,* volume 1: *Ecological Vision and Theory for Temperate Climate Permaculture.* I couldn't believe it was happening (or how long the book was!). Though I had written little of volume one, every page of it was certainly stamped in my brain from countless edits and reviews, and there in the back was my piece on the top one hundred species for forest gardens, including many of my photos from our garden in Southampton.

It's an emotional moment when your first book comes out. I had been a writer for nine years, but now I was an author. I could not control my grin for the next several days. I can't remember when I first saw volume two (*Design and Practice*), even though that is where the bulk of my work lay. I do remember returning many times to see in print the designs that Jonathan and I had worked so hard on. It felt as if we had put out a challenge to the world, and it was time for us to deliver what we had promised.

There were signs that this might actually happen. A local friend and permaculture teacher named Ethan Roland had read *Edible Forest Gardens* and *Perennial Vegetables* and started to come by every couple weeks to see what we were doing, how things were growing, sample new foods, and talk about the practices and our movement. He had never eaten many of the perennial vegetables we were growing accustomed to (few people in North America had), and we were excited to try new berries together for the first time.

One day another permaculturist named Justin West visited. Justin was from New Jersey but had been working closely with Martin Crawford in England at the forest garden that is still to my knowledge the largest and best forest garden installation anywhere in cold climates of the world. Dave and I had toured Martin's garden in 1997 as research for our book, and I had fantasized about studying with Martin. Now here was his protégé coming to check us out.

After a long afternoon of wandering through the garden, our conversation a geek delight, we prepared a dinner featuring a range of spring perennial vegetables. In particular our fuki stalks in ginger-vinegar sauce were a huge hit. The visits from Ethan and Justin marked a turning point in my life. Up until then, I had been the young permaculturist traveling to see established sites. Now I had finally planted a garden the others wanted to come and learn from. Though it was still young, its promise was enough to excite the up-and-coming leaders in my world.

Jonathan and I sat under our newly erected kiwi trellis with Ethan, Justin, and his sweetheart, talking long into the night about Martin's garden and the edible ecosystem emerging before our eyes. As the sun set, the city lights came on, and we could make out the forms of our annuals, herbs, and shrubs blowing gently in the evening breeze. It was a long time before I felt ready to pick up our dishes and move the action inside.

For the first time, I felt completely confident that what Jonathan and I were doing was going to work. For the next several years,

Justin would visit us on his return trips to the United States, serving as a human dispersal vector and cross-pollinator of ideas from our young garden to Martin's far more established operation. It felt like the beginning of a new phase in the permaculture movement in the United States. At this point British forest gardens were ahead of the game by a decade or more. We knew few such projects in the United States, and none with the fully fleshed out understory and broad selection of perennial vegetables, nitrogen-fixers, and insect nectary plants that were beginning to thrive in our urban lot.

PART THREE

LEAP

2007–2009

17

EXCESS SUCCESS

Ornamental perennial gardeners know that it takes time for perennial plants to get established. The first year perennials may not appear to grow much at all. The second often brings only slow growth. But in the third year, with deep roots anchored in the soil, they explode. The phenomenon is known as "sleep, creep, leap." To this Dave Jacke adds "reap," because in our case we're talking about food-bearing perennials.

When a whole garden full of plants hits "leap" at the same time, something happens at the level of the garden ecosystem as well. By 2007 it was clear that our garden was coming to life. Part of this must have been because of our gradually improving soils and the habitat we'd created for beneficial insects and other helpful organisms. Our shrubs and perennials were enjoying riotous growth, and many of our young trees showed promise.

The years when annuals dominated our garden were over. The ground was carpeted with a dense growth of perennials whose foliage made a chaotic pattern of different colors, leaf shapes, and textures. Feathery purple Ravenswing wild chervil grew next to light green, woolly Turkish rocket and gray-blue sea kale. Deep-green groundnut vines climbed everywhere, and the sprawling branches of nonedible, nitrogen-fixing licorice milk vetch intermingled with the foliage of edible species like garlic chives and good King Henry. We had to trim along our pathways just to be able to get in to harvest. (This milestone has long since lost its charm.)

Many of the plants we introduced actually started to become weeds themselves. Anise hyssop, garlic chives, sweet cicely, and licorice milk vetch, all of which we had lovingly nurtured from tiny seedlings, went feral—seemingly overnight. The worst was a beautiful vine with the unwieldy Latin name *Thladiantha dubia*. In our search for a good-quality hardy perennial cucurbit—a perennial counterpart to winter squash, zucchini, cucumber, or watermelon—Jonathan and I begged for a few *Thladiantha* shoots from a friend. *Thladiantha* is a climber with yellow flowers that are attractive to hummingbirds; their young shoots can be cooked and eaten, though they are not outstanding. The plant produces small, red, juicy cucumbers, but we never actually saw this for ourselves since our plants were all male, and our efforts to find them female mates were (thankfully) unsuccessful. When we noticed shoots coming up more than ten feet away from the pampered parent plants, the species earned the nickname Thlad the Impaler for its aggressive underground spread and sprawling growth, and we abandoned our search for female mates and began to concentrate on eliminating this troublesome disappointment.

Neither Jonathan nor I have ever been terribly excited about weeding. It doesn't fit into our low-maintenance fantasy. But there we were with a problem of excessive vegetative growth. In permaculture, when something in your garden starts bugging you, it means your garden is trying to tell you something. Ours was trying to tell us, rather loudly: Put me to use! We had to learn how to harness this newfound abundance and turn it into something productive instead of battling it by weeding for the rest of our lives. We came up with three strategies that were to profoundly change our interaction with the garden for the better.

Our first problem was that we had an excess of prunings and woody stalks from sunflowers, corn, and shrubs. We'd purchased an electric chipper, but it was a failure, and we never liked the idea of using electricity or petroleum to manage our garden anyway. But Jonathan's sweetheart, Megan, knew right away what to do with all of that material. She purchased an outdoor portable fireplace and turned our piles of prunings into s'mores and fun nights around a fire. And we could use the ashes in the garden. First problem solved.

Other weeds arrived entirely on their own. Straw mulch always seems to bring oats, various weeds came in on the compost, and others just seemed to appear magically once we had improved our soil. Our compost piles were becoming heaps of weeds and excess crop foliage and did not break down satisfactorily. Between our weeds and our riotous crop growth, we needed a third solution to rapidly convert lots of fresh leaves into compost.

18

TURNING WEEDS INTO EGGS

Marikler and I had a small wedding in the bamboo grove in our backyard in July 2007 and a larger ceremony with friends and family that fall. (Following Guatemalan tradition, Marikler wouldn't move in until we were married.) Not long after our wedding, I woke up one morning with the strangest feeling. I couldn't place it at first and then realized the headache I'd had for fourteen years was gone. It had slowly been getting better, but until then I had had one very long headache ranging from minor annoyance to brain-crushing migraine. I was freed. I think it was being in love with Marikler that did it.

Jonathan's sweetie, Megan, moved in shortly after Marikler and I were married. He and I had been living on one side of the house and renting the other, but now we each moved to our own side of the house and our beloveds moved in with us, only one month apart—a most unlikely scenario. It was the beginning of a new chapter in our lives. We felt like two birds that had built nests full of shiny objects and successfully attracted mates. Surely Marikler and Megan didn't fall for us just because of our garden of fruits and flowers. But the garden was an extension of who we were. And the vision of a life with golden raspberries and tender winter salads certainly didn't hurt our prospects.

After Meg and Marikler moved in, we began to joke about the house being a two-unit ecovillage or a very small condominium complex with an awesome backyard. One of our first decisions as a "homeowners' association" was about chickens. Jonathan and I knew that chickens would help to us turn our excess vegetation into fertile soil and eggs, but we were intimidated by the idea of breaking the law. Neither of us had raised our own layers before, though we had raised one round of meat birds at Wonder Bread. Both Meg and Marikler, on the other hand, grew up with chickens: Meg on a farm in upstate New York and Marikler in a small urban backyard outside of Guatemala City. Their enthusiasm won us over, and we decided to undertake chicken raising as an act of civil disobedience. We knew our efforts would need to be neighbor-friendly, so we chose to have a small number of birds and no rooster and to make every effort to run a clean and sanitary poultry operation, with no smells that might bother our neighbors.

One day in March 2008, our chicks arrived in the mail. Marikler and I set up a brooder in our basement with a heat lamp, waterers and feeders, and bedding. We had to dip the chicks' beaks into water so they would learn how to drink, and they needed encouragement to learn to eat their feed as well. They got their water dirty almost the instant we set it inside their enclosure, and we had to keep the heat lamp hanging at just the right height so they could all cluster underneath, but without setting anything on fire. We were constantly adding more shavings to their bedding, as a layer of chicken manure seemed to build up every time we went upstairs for a few hours. Somehow the whole operation generated dust, which formed a thin layer on every surface in our basement.

But those little puffballs also brought us a lot of joy that spring. From the first week they also began to help us with our weed problem. Though there was still snow on the ground outside, miner's lettuce had already become a little too luxuriant in our greenhouse. We began to feed the leaves to our chicks. Still, it's a good thing they were so adorable and we went down into the basement all the time to visit them, or the chores would have been a major effort.

We added a small chicken house to our toolshed and built a protected run outside to defend our girls from cats, hawks, skunks, raccoons, and the occasional roving pit bull. We partitioned off a section of the interior of the shed for their shelter and added a fenced-in and netted run outside for the straw yard. It seemed a little excessive, but nothing has ever got in and messed with our girls.

By the time we moved the young birds outside, our first crop of outdoor weeds was coming in. Our straw-mulched beds were growing weedy oat greens, which make wonderful chicken feed. Whenever we brought them fresh greens, our girls would run to the door of their chicken run and cluck in excitement. Our weeds became scarce, and weeding went from being an arduous chore to a delightful pastime. Since then, we've never had a serious weed problem.

But of course chickens do more than eat leaves and weeds. All of the weeds and excess crop vegetation that we throw into their run is picked at, scratched up, and pooped on. By the end of the growing season, manure accumulates to a depth of about five inches throughout the run. We often apply this directly to established perennials. Sometimes we put it in a pile and cook it for a few weeks before spreading it on annual beds. Chickens have tightened our fertility cycle and improved productivity. Where once our excess green leaves stagnated in compost piles that wouldn't heat up, the chickens shred them and mix them with manure and straw by scratching, speeding up the efficiency of our composting enormously—we don't have to turn compost ourselves anymore! Eggs were also a welcome addition to our household diet. We learned to explore the seemingly endless variations of eggs with fresh vegetables; fried rice, frittatas, scrambled eggs, and omelets graced our tables.

Everything was going fine with the chickens until one late summer day in 2008. I came home from a long day at the farm and was puttering around in the backyard, feeding a few leaves to the chickens when a blur of dark blue motion caught my eye. Our neighbors' yard was full of police, guns drawn, climbing quietly up the stairs to the third-floor apartment. It looked like a scene from a ten p.m. cop show. My next, more practical thought was that I should probably get back inside the house right away so I didn't get shot in a firefight.

For the next two hours, our neighbors' yard was crawling with police busting an eccentric and stylish tenant who was selling cocaine.

He had once kept a beautiful horse in their small backyard for a few weeks and was known to leave pig heads lying around after a pork roast. When Jonathan, Meg, and Marikler came home from work, we worried that somebody would call in our chickens. While backyard poultry hardly compares to coke dealing, I knew how the powers that be in Holyoke felt about our crime.

Sure enough, within an hour the red pickup truck of our animal control officer pulled up, blocking the cars in our driveway. He marched into the backyard, giving our operation a thorough inspection and looking like he was prepared to shut it down. I think he was actually pleasantly surprised by what a nice job we were doing caring for our birds, since a lot of his job involves busting cockfighting operations. He told us we needed to get rid of our chickens, gave us a written warning, and said that if we got a complaint again we would be fined. Later some of the junior officers told us across the fence that it was their captain who had called in the chickens. They, on the other hand, thought that the chickens were cute and were happy to see them there.

What to do about our girls? We were so accustomed to our chickens it was hard to imagine doing without the small herbivores to manage our garden fertility cycle. We found someone in the country who was willing to give our chickens a place to lie low for a few weeks until the heat cooled off. Bringing the birds back at all seemed risky to me; I was sure we would get into terrible trouble if we were caught again, until Meg and Marikler pointed out that the penalty for another complaint was a twenty-five-dollar fine, something we could probably live with. It helps that most of the neighbors who abut our yard grew up in Puerto Rico, the former Soviet Union, or Brazil, where urban poultry are part of the norm. "Oh, well," we thought. "If they shut down our chickens, we can always get rabbits." In Holyoke rabbits are legal because they are "pets," and though rabbits don't lay eggs, they do taste delicious at the end of the season.

That fall of 2009, legislation was introduced to the Holyoke City Council to legalize backyard chickens. It would have allowed a small number of hens and made available a limited number of paid permits. The proposed measure ignited a firestorm of controversy; letters to the editor in the newspaper opposed "taking Holyoke backwards." The chief of police ridiculed the idea. There is a legitimate concern in Holyoke about cockfighting, but hens have nothing to do with this. Rather, poultry seems like a racially charged issue. Most of the people who keep poultry in town do so as an expression of their Puerto Rican heritage. Most of those who oppose them are white and left that part of their history behind generations ago. Despite a reasoned letter from our town's beloved veterinarian, arguing for the health benefits of backyard poultry, the measure failed.

I'm not proud to admit that Jonathan, Meg, Marikler, and I were silent in this debate. We felt we needed to keep our heads down so as to avoid being the proverbial spouting whale that gets the harpoon. We did do some research and found that many of the surrounding towns had recently decided to allow small numbers of backyard hens. Most surprising was to learn that in Manhattan one can have up to fifty birds.

Although we would have benefited from an ordinance allowing for a small poultry flock, I didn't think it was the right time for the legislation to be brought forward. There were other policy recommendations championed by the Food and Fitness Policy Council (a coalition working to prevent youth obesity and diabetes that I was part of through Nuestras Raíces) that had the potential to make a bigger difference to the health and sustainability of the city—measures to increase redemption of WIC farmers' market coupons, increase the number of community gardens in new neighborhoods, and create farm stands and mobile vegetable sellers.

Chickens can't live on greens alone. We give ours bagged feed every day, although what they really like to eat is bugs. Every time we roll over a log and find an ant nest, we grab a shovel and a bucket and feed the wriggling larvae to our chickens. Aphid-covered leaves, pill bug colonies, and even slugs contribute to their fare.

A few years ago, when our friend and fellow permaculture designer Ethan Roland offered me baby silkworms, I leapt at the opportunity to incorporate them into our growing systems—that and I needed a birthday present for Marikler. When I saw the delight on Marikler's face as I presented her with a box of pallid, wriggling caterpillars, I knew I'd married the right woman.

Silkworms are easy to raise. They eat nothing but mulberry leaves, and at the time Ethan gave us the silkworms, our mulberry was growing excessively. We keep the silkworms in a cardboard box, feeding them fresh leaves twice a day. When there get to be too many worms, which are full of fat, protein, and calcium, we feed some to the chickens. By the time they reach about two inches long, they are mostly made of silk and lose their food value for chickens. We try to let about twenty of them survive to this stage. They spin beautiful cocoons of silk, some of ours producing white silk, others golden. In commercial silk production the pupae are boiled alive inside their cocoons, and the silk is drawn off of the pot. The boiled pupae are fed to livestock or sometimes eaten by people. (Marikler tried one during travels in Southeast Asia and told me it tasted sweet; somehow I have not got around to eating silkworms yet.) Since we're not harvesting the silk, white moths emerge from our cocoons after a week or two, their antennae looking like giant, delicate eyelashes. Silkworm moths are flightless, the result of centuries of domestication, so there is no risk that they will escape into the environment and become a new pest. They mate, lay eggs, and die in a short period. The eggs look a lot like arugula seeds and can be stored in the refrigerator over winter until the next mulberry leaves are ready. If we had more mulberries, we could hatch out two to three hundred silkworms every week and provide a significant component of our chickens' diets. At our current scale the worms are an occasional treat.

I knew when I married Marikler that her main love was livestock, not plants. So I was surprised at how much time she spent in the garden. Gradually I understood that it wasn't the plants she was observing but rather the animal life.

For a long time, Jonathan and I thought of wildlife in the garden only as a distraction. Our garden was for plants; we judged animals solely on their positive or negative impact on food production. Marikler was constantly pointing out the antics of squirrels and sparrows, animals Jonathan and I considered annoyances or vermin. But above all she was fascinated with insects and spiders. Using Whitney Cranshaw's *Garden Insects of North America*, Marikler learned to identify insects at many stages of their life cycle. "Look," she would point out in the garden, "Here are ladybug eggs. See how golden they are? And here you can see their larvae eating aphids."

Once, after weeks of watching a female praying mantis and following her around the garden, Marikler showed me something remarkable. Our female praying mantis looked like she was spraying shaving cream out of her abdomen. She was making her egg case. Each fall Marikler notes the location of praying mantis egg cases around the garden so we can watch the tiny young mantises emerge in spring.

Marikler's sense of wonder helped me realize that our garden was not just a place for food production but a place where life happens. I knew that intellectually, but it never fully sunk in until I met her. Jonathan and I had put into motion a process to restore a piece of the planet, and creatures were taking notice.

19

GRAZING BERRIES

During his regular visits, our friend Ethan Roland spent many hours walking in the garden with us, making observations, and sampling perennial produce. In the middle of one of our grazing tours of the garden, as we moved from one species of berry to another to another, Ethan observed that not only were we eating a high diversity of tasty berries but, much more unlikely, our sweeties had both moved in within a month of each other. Jonathan and I decided we should review our goals to see what else might be coming our way.

As perennial vegetable season is drying up, berries are coming into full swing. Foraging for fresh fruit in the backyard was a key goal for Jonathan and me, as reflected in the diversity and abundance of berries in our garden. Within two to three years, all of our berries were yielding well, and many were filling in to form nice patches. There's nothing better than walking out the back door and feasting on five or six different kinds of berries as you make your way through the garden. Jonathan and Meg love them on their cereal every morning, and we've cooked all kinds of dishes with them. But in general that's too much work for me: nothing is as satisfying as filling up a handful of berries and shoving them into my mouth right out in the sunshine.

Our first berries of spring are honeyberries, or haskap, which ripen in May. You need at least two varieties for pollination, and though our first one died, its replacement eventually caught up. We enjoy these small, sweet-and-sour blue delights, a harbinger of harvests to come. Soon after the honeyberry harvest is over, strawberries come along. I never liked strawberries growing up; now I know it's because the quality of strawberries sold in stores is nothing like a fresh, fragrant strawberry from the garden. We rotate our strawberry beds slowly through the garden, giving each planting three years in each location and starting a new one every spring. We started with everbearing varieties but then turned our focus to heavy June-bearing types, since we have plenty of other fruit during other times of the year. Alpine strawberries, which pack all the flavor of a quart of strawberries into each tiny fruit, start fruiting around the time, continue through early July, and come on again to some degree in the fall. They're so small and make just a few fruits each day, so we plant them along the borders of our pathways to remind us to take advantage of them.

Some years our strawberry harvest is poor. Perhaps it is cold and wet, and disease is affecting our plants, or perhaps we didn't do a good job of weeding or planting out our new bed the year before and yields suffer as a result. In those seasons we rely heavily on our goumi, whose small, red cherries are tart and astringent until dead ripe, at which point they're quite nice. Our goumi bears heavily, so we learned to process them with our steam juicer to take full advantage of the harvest. Of course goumi is also welcome in our garden because it is a nitrogen-fixer and grows (and bears) extremely vigorously in some of our worst compacted clay soils and in partial shade.

Later in June we get a second flush of berry species, such as those from our Gerardi dwarf mulberry. Most wild mulberries are watery and insipid, but good cultivated varieties have firmness and tartness in with the sweet. Our dwarf mulberry bush bears large, tasty fruit for about six weeks. It's not quite as dwarf as we had thought, and we spend a lot of time pruning it back in winter and keeping it under control in summer by lopping off branches to feed to our silkworms. Left to its own devices, it might spread to ten feet high and wide.

It wouldn't be June without juneberries (also known as service-berry, sarvisberry, saskatoon, and shadblow, among other names). We planted our two Regent saskatoons, a cross between a treelike and a dwarf variety that reaches an ideal five-foot size, in a prime location between the greenhouse and the beach plums, a spot with ideal sun

and our least terrible soil. Regent bears heavily, and like all juneberries has knockout flowers in spring. Little did we know that in our
garden the fruits of this variety would be somewhat dry and mealy
(like many of those I've eaten in the East, though the ones I've had in
Colorado are sweet and juicy). Jonathan and I ate them out of a sense
of duty, but Meg and Marikler never felt so constrained, and the fruits
were just not getting used. Even the birds were not enthusiastic. We
moved the Regents to the front yard a few years ago so that at least
their pretty flowers would be visible to passersby and the neighborhood children can eat all those berries they want. Much more to our
liking is a wild clone of running juneberry, which we dug up from
a nearby natural area a few years ago and transplanted in with our
blueberries. The berries are a little smaller but sweet and juicy. This
wild juneberry doesn't grow more than two feet tall, but it spreads by
runners and is filling in the gaps between our half-high blueberries.

Toward the end of June, our red and white currants come in. We
found some clones (Red Lake, Blanca, and Pink Champagne) that
are nice for eating raw. At first I was the only one in the house who
liked them, but once I passed on my secret of lightly chewing the
fruit but leaving the seeds intact, the others started to enjoy them as
well. When Megan cooked down the red currants to make a sauce
that she, Jonathan, and Marikler had over ice cream, they were truly
hooked. Since then we have planted more varieties. Red and white
currants bear well even in full shade, which means they're taking on
an increasingly prominent role beneath our mimosa and fruit trees as
they mature.

July brings ridiculous riches of berries, many belonging, like currants, to the genus *Ribes*. Gooseberries are a house favorite. Most of
the varieties we grow have spines, making the harvest quite painful,
but their flavor is like a grape from another dimension. We grow
about six varieties and enjoy all of them for fresh eating when ripe and
in baked desserts when still green, tart, and firm.

Black currants ripen in July as well. These were an acquired
taste for me, though I enjoyed beverages and desserts made with

them many times. In 1997, when I traveled to England with Dave to research *Edible Forest Gardens*, we stayed with Robert Hart, who literally wrote the book on temperate climate forest gardening. At the time, a local bakery would trade Robert his black currants for their desserts. I ate black currant crumble for several meals. When fully ripened, black currants' musk and spiciness is emboldened by a sweet and juicy essence. I nibble on them as I walk by them in my own garden, though I don't eat a double handful the way I do with red and white currants.

Black currants and gooseberries were crossed to create the jostaberry. Jostas are happy in our garden—even in pretty serious shade under Norway maples—and produce well for us, plus they are thornless. Though jostas are tart until fully ripe, I can't get enough of these spicy, sour delights.

Our final member of the genus *Ribes* is clove currant, which has clove-scented yellow flowers that bloom around the same time as forsythia; unlike forsythia, however, the flowers are followed by edible fruit. Our Crandall clove currant has fruits that are much larger than wild clove currants, though I think their skin is tougher, and I prefer the smaller wild forms. Jonathan says Crandall is his favorite fruit, and the neighbors love them as well.

Our clove currants grow on a narrow strip of terrible soil between our driveway and our neighbors' driveway, an area that is hot and dry in the summer and buried under several feet of snow in the winter. Our goal is to create an edible hedge there. This is where we moved the Regent juneberries that we weren't totally in love with, and where we also fruited a sand cherry, a species native from Cape Cod through the Rocky Mountains. We grow two prostrate forms (Pawnee Butte and Select Spreader) that serve as groundcovers. The black fruits ripen in July and have a rich, subtle flavor. I look forward to their fruiting more heavily in the future.

Blueberries are next in July's carnival of abundance. We grow a couple of hybrid half-high varieties and some lowbush blueberries as well. So far we have not had to net them the way our rural friends do.

Birds in the country will eat all the berries before they even ripen, but our birds in the city don't seem to know what to do with them. Our blueberries have grown slowly, and we have messed around with the pH several times trying to give them a jump-start. They eventually began to bear pretty well, though Megan and Marikler still go to a pick-your-own operation to get enough berries to freeze for winter. Generally, I'm pleased with our strategy of planting small numbers of a great diversity of fruits to have the longest possible season, but when it comes to blueberries, I wish we had a second tenth of an acre to dedicate to them.

When we moved in, we noticed right away that there was a patch of feral raspberries behind our neighbors' shed. The branches that hung over the fence gave us the first fruits we ate in our garden. The berries were small and sweet, with a delicate flavor. We later purchased and planted a fancy cultivar and were disappointed to find that the fruits, though large, were tough and bland. We ripped them out and transplanted in some of our neighbors' wild plants. But our favorite raspberry is the golden Anne, which I'd ordered from a nursery after tasting it in several gardens and falling in love. Jonathan calls them "mango berries." I agree that they taste fantastic, but I also love the insight they offer into an elegant pest-control strategy. Birds can be pests in raspberry patches, but they wait for the fruit to turn red to know that it is ripe. Yellow fruit doesn't register on their radar. This also applies to yellow cherries and perhaps other fruits and represents an interesting alternative to netting your fruit.

The berry season peters out for a while after raspberries, so we plan to add black raspberries, thimbleberries, and thornless blackberries to help fill in the sad two weeks of the summer when, in a bad grape year, we sometimes have almost no fresh fruit.

Raspberries return again for a halfhearted season in September and October. Late September is also the ripening time for one of my favorite little-known fruits: ground-cherries, sweet relatives of tomatillos and tomatoes. A papery husk encloses a golden orb, and the flavor is often compared to a combination of pineapple and cherry tomato. We grew annual ground-cherries for several years but over time switched to the native perennial species like clammy and longleaf ground-cherry, which have smaller, firmer fruit with the more complex flavor. I have picked ripe fruit off the plants as late as December 21.

I'm not sure if wintergreen berries are our first or last fruit of the year. They ripen in late fall and remain viable under the snow all winter. Wild forms have tiny fruits, but I noticed a number of years ago that cultivated ornamental forms like Very Berry and Christmas that have been selected for larger flowers also have much larger fruit, more of it, and share the same wintergreen flavor as the wild forms. Marikler thinks the berries taste like Pepto-Bismol, but I like them.

Our late-berry season also includes a bit of lingonberry fruit from a six-inch shrub related to the cranberry. The fruits are like tiny cranberries, though we have not yet had enough to do much with them.

Overview, 2004. Turning our backyard into a healthy, edible ecosystem was going to be a challenge.

Sheet mulching in spring 2004. One of a series of work parties that made our garden possible.

Overview, spring 2007.

Main underground tuber and aerial bulbils of
the edible perennial Chinese yam.

Tropical garden at the front
of the house.

Marikler and I were married July 7, 2007, in the bamboo grove.
Photo by Benjamin Jacques.

Meg and Jonathan enjoying the harvest.

Overview, spring 2009.

Chinese lotus in our edible
water garden.

By 2008 our perennial vegetables
and other herbs were growing
almost too vigorously.

Our edible forest garden in 2009,
with clear tree, shrub, and herb
layers emerging.

July is berry season. Clockwise from top: red raspberry, white currant, goumi, red currant, blueberry, mulberry, jostaberry, yellow raspberry, gooseberry (at center).

Pawpaw, a luscious but underutilized native fruit.

A blue salamander that showed up on its own in 2009, indicating improving ecosystem health.

Marikler feeds excess leaves to the hens and their chicks.

Jonathan with delicious wine cap mushrooms grown in wood chip mulch.

Overview, 2011.

A next-generation polyculture: sunchoke and hog peanut.

Jonathan leads a tour in the front garden.

Our outdoor living room has become a quite private retreat in the city.

The Snowtober 2011 storm crushes our greenhouse.

Our bioshelter greenhouse under construction in early spring 2012.

SECRETS OF RESILIENCE
by Jonathan Bates

CATCH WATER

There is a blue, fifty-five-gallon drum half-filled with water at the edge of the garden. The same clear, clean water has been sitting in the drum for seven years. Perhaps this sounds odd to you. But along with the soil on which it sits, it is an important part of the story of water in our garden.

Our roof is also a key part of that story. Its main function is to shed water away from our lives; without it our home's foundation would be ruined by rain and would disintegrate from rot and rust. Most cities and towns encourage home design to point roof water to the street. In our town the water washes to an energy- and chemical-intensive sewage treatment plant. Is this the best use for rainwater?

One spring day I watched a downpour of rain collect and careen down the neighbor's roof, puddling in our yard, some soaking into the ground, a lot more running away. I thought that with some simple materials I could redirect the water from my neighbor's roof and use it in our garden. Within weeks I had what I needed: old plastic gutters that I found in someone's trash and a fifty-five-gallon drum that I picked up at a local salsa factory.

After getting permission from the neighbor, I fastened a gutter to some very tall stilts that leaned against the back of the neighbor's garage wall (which sits directly on our property line). I positioned it to catch the water running off the roof, sending it to the barrel via a gutter and downspout. No other attachments were necessary.

The invention lasted a few months, until our neighbor sold his house to a new owner who dismantled my gutter and allowed the water to flow again onto the ground. The last half-barrel of clean water we collected and never used sits like a monument to the simplicity of water catchment and storage in our garden, inspiring us to do more. We now have additional rain barrels, including two we got free from the city. We're planning to use them to collect water from our greenhouse and shed.

Along with collecting water, we also conserve water by covering the soil, building thick, rich, mulched garden beds that stay moist longer. The mulch suppresses weeds and as it rots down builds soil. Imagine hard, compacted ground—soil that is so pushed down that the air pockets making it spongy are gone. Water runs off compacted ground. But if you aerate it and make it light and fluffy, water passes into it very easily, allowing plant roots to breath and flourish. Over time earthworms have multiplied and turned our garden beds into little hills of crumbly castings, filled with a diversity of life, stabilizing and enhancing soil chemistry and transforming the soil into a sort of water battery.

I try to remember that water is a borrowed element. It has been moving through our environment and our bodies for billions of years. We can either ignore how humans are part of this cycle and flush water away, or we can slow it, spread it, and sink it—using it as it moves through our landscapes and lives. We'll continue to borrow water, turning a house rotting, soil-eroding water problem into a life-sustaining, resilient water solution.

BUILD SOIL

With wood chip mulch, rotting logs, tree prunings, leaf litter, organic compost, chicken manure, and the organisms and forces interconnected into a web of soil decomposition and growth, we're able to raise the food we eat. When designing our garden, I for one didn't know a lot about this soil-building process. I knew at a basic level that soil was an important component of growing things, but I had to learn how absolutely critical healthy soil is. It took living in a garden, seeing for myself the process of life and death and rebirth in the soil and garden, for me to start understanding the interconnectedness of it all.

We had very little good soil when we moved into the house in Holyoke. What worked more effectively were four components: deep aeration, on-site manure, good compost, and micronutrients.

Deep Aeration

In year six, with a clearer understanding of the benefits of noncompacted soil, we decided to do some deep aeration with a broadfork. How the heck does one do this in an already established perennial garden? Not easily. We used a heavy-duty, all-steel Meadow Creature broadfork, one that can go down at least twelve inches and can stand up to compacted clay soils and urbanite. To use it, you stand on the fork's U-shaped bar, hands on the vertical handles at shoulder height, pushing the tines into the ground with your weight. Once your feet and bar touch the ground, you rock the bar back, from upright to forty-five degrees, lifting the soil to aerate it but not turning it (which would disturb the complex, diverse community of soil microorganisms). Then you pull the fork up and out, stepping back six inches, and repeat the process. Deep aeration may have been the most important advancement we made in regenerating our soil and growing healthy plants.

On-site Manure

Before the garden was established, I experimented with growing earthworms in a container. The process was as simple as putting a pound of red wigglers along with damp, shredded newspaper into a plastic tote with holes drilled in the top for air. Every couple of days I would feed them vegetable scraps and watch to make sure they were happy (not too wet and overfed, both of which causes anaerobic conditions). This little experiment, which anyone can do under the kitchen sink, was very successful. The process produces wonderful soil I fed to my houseplants. To my amazement, a similar worm-created soil was building in our garden. This process was happening so well and in such great abundance that in many places in the garden I could pull back the mulch, shovel my hands into the ground, and get rich worm manure by the bucketload. Under the right conditions, of soil moisture, organic mulch, and generous plant and soil biodiversity, the garden can harbor uncountable masses of worms. During a clear quite night I can go into the garden and hear them retract into their holes as I walk past.

Our chickens became a much-needed source of nitrogen-rich manure, too. We've found that by putting their old bedding right into the garden, onto empty annual beds, for example, we can compost in place and plant vegetables right into it the next season.

If our local regulations permitted it, we would close the cycle by adding human manure (along with the urine we already use). Some day, if someone is so bold as to demand it from our politicians, municipal-composted humanure might be the norm.

Good Compost

Although we compost some of our vegetable kitchen scraps in a prefabricated plastic composter, we don't generate enough bulk compost from home production to add to our vegetable beds each year, so we do have to buy compost to meet our needs.

Purchased compost is not created equal. We've tried many different vendors and bagged varieties, too. Eventually we found a local dairy farmer who makes rich, organic compost that has the right ratios of grit and organic matter, holds water well but doesn't get too boggy or dry out, and is made with local waste materials like cow manure, supermarket vegetable waste, and shredded leaves. This compost enables us to grow the best organic vegetables in the world!

Micronutrients

Micronutrients in our garden come from sources such as dead wood (and are accumulated by mushroom mycorrhizae), manures, rock dust, subsoil or bedrock from deep-rooted plants and their leaf mulch, purchased organic dry amendments such as chicken manure, blood meal, rock phosphate, greensand, bonemeal, and purchased and brewed nutrient sprays such as compost tea, fish and kelp emulsion, humates, molasses, and sea minerals. Over the years Eric and

I have used many of these, and all have helped in different ways.

The most effective ingredient to healthy soil creation is usually the simplest option. During my early garden learning, a mentor turned to me and asked, "Why are you picking the stones out of the garden beds?"

"I always thought that's what you're supposed to do," I replied.

"Gardens need minerals, right?" he responded. "If you pick all the rocks out, where are the plants going to get the minerals?"

Since that day, unless the rocks are in the way, I leave them in the garden.

INVITE DIVERSITY

While in the garden one summer, observing the whirling cloud of bees, wasps, and flies hovering and buzzing around flowers, I saw a plump, green stick of a bug, something out of a horror flick or tropical rain forest—the wonderfully skillful hunter at the top of the insect kingdom: the praying mantis. There it sat, waiting for the perfect moment to pounce. A slow twist of the head and snap! Its long, needle-lined legs grabbed dinner—a honeybee, no doubt full of sweet flower nectar.

Along with praying mantises, reptiles are also har-bingers of ecosystem health. Since our neighborhood had so many cats around and little connection to other "wild" areas, initially I didn't think that snakes could find their way to our ecological island. But while going through old flowerpots one spring, I saw a little brown snake curled up in the bottom of a pot. There was still chill in the air, so it hadn't woken up yet. I reached down and picked up the little fella and carefully placed him into the deep mulch nearby. What an exciting find! The brown snake is an avid slug hunter, which makes it a perfect addition to our garden.

20

FRUITS AND NUTS

Since we have little room for fruit and nut trees, we had to prioritize the species we most love to eat, with the prime fruit-growing space going mostly to underutilized native fruits. Still, the August fruit and nut season is something we look forward to every year.

In dry years we have phenomenal yields of grapes. One epic year we had about ninety bunches. In a year with typical rainfall, we harvest more like twenty to thirty good bunches and some additional partial bunches. We have trellised our vines on a wrought iron fence with a little bit of additional structure strapped on to keep from having to build something fancier. Our grapes are hybrids of the native labrusca Concord type and European vinifera grapes. We selected varieties—all of them seedless—using a disease-resistance chart in Lon Rombaugh's *The Grape Grower*. We had a splendid Interlaken green that perished one winter, although our pink Glenora and purple Reliance soldier on. I don't think I had ever had a really fantastic fresh grape until our three vines began to bear.

Most seasons we battle fruit rot with some cultural controls. For a few years early on, we sprayed our grapes with hydrogen peroxide, but we had to begin doing it as soon as the flowers bloomed and are just not dedicated enough to spray anything consistently, let along perform repeated applications for a month. The strategy we settled on instead is to prune away all the leaves around the fruit so that sun and breeze keep them as dry as possible and minimize fungal and bacterial growth. Grapes are the only fruit we grow that we put this much work into, but we think they're worth the effort.

Coming in just after the first grapes are our beach plums, a native species that can be found along the coast from Maine to New Jersey. The wild forms can be astringent, but our improved varieties make for nice fresh eating. But it is as jam that beach plums come into their own. After one of our beach plums was hit by a peach borer and killed back to the ground, it resprouted, but not vigorously, and as a result we have not had excellent pollination since then. During the few years when everything has gone right, however, the branches of the plum are loaded with fruit and are a testament to the potential of domesticating native wild edibles.

One of the biggest drawbacks of the house we chose was that there was no room to grow chestnuts, walnuts, or other large-sized nuts that Jonathan is such a big fan of. Hazelnuts are an exception. We have two dwarf hazels that took some time but eventually began bearing decently. Though the nuts are somewhat small, they represent one of the best forms of perennial protein we can grow in our garden.

Toward the tail end of August, we start eyeing our Asian pears to see when the fruit will ripen. Since we have limited space and Asian pears require a pollinator, we splurged on a semidwarf tree with three varieties grafted on it. All three varieties flower at the same time and pollinate each other, but one of the varieties is early-, one is mid-, and the third is late-season ripening. This strategy maximizes fruit in a small area. We have kept the tree to about ten feet high with pruning, and it is no more than eight feet across, but it yields well. As with our grapes, we have to do a bit of work for our Asian pear tree. Every spring we face the heartbreaking task of thinning the fruit by hand. It's so difficult to remove a fruit that will someday become an Asian pear, but it pays to encourage a yield of fewer but larger fruits.

We lost the tags and no longer remember the names of the varieties. The earliest, though small, green, and sweet but bland, is certainly welcome in the last week of August, when few other fruits are bearing. The second variety, which I believe is Chojuro, is crisp, juicy, and has a caramel flavor and russet skin; this is a sublime contributor to September harvests. Our small tree produces perhaps fifty or sixty fruits of this variety per year; it would be nice if there were more. We are still getting the hang of pruning a tree with three different genetic individuals grafted to it, and our late-season type has suffered as a result. This variety, which may be Korean Giant, puts on vigorous vegetative growth and is crisp and sweet but doesn't seem to fruit much. It grows more rapidly than the other grafted members of its shared tree, so we have pruned hard in subsequent years so that it doesn't steal all the energy. Unfortunately, this has meant that we get only about a handful of pears from it every year.

October brings a moment we wait for all year: pawpaw harvest. In the spring the brick-red pawpaw flowers open, but honeybees are not interested in them, as their fetid odor is intended to attract carrion insects like flies and beetles. This is a vestige of the early days of pollination: the Magnolia order of plants, to which pawpaws belong, evolved before the origin of honeybees. Some commercial pawpaw growers hang roadkill or buckets of fish guts in their trees to improve fruit set. Needless to say, this is not an appropriate strategy in our neighborhood. We have made some effort to plant other carrion-insect-pollinated species around and beneath our pawpaws, but we always hand-pollinate to ensure a good harvest.

Gender in plants is complex and interesting. Flowers can be male, female, or perfect, meaning they have both male and female reproductive organs. An individual fruit tree might have one, two, or even all three of these flower types, which does not necessarily correlate with whether or not a tree pollinates itself. In the case of pawpaw, each flower begins female and then becomes male, but the trees are incapable of pollinating themselves. So every spring, when the first flowers have become male and are shedding pollen, Jonathan and I head into the garden with a tiny watercolor brush and a little dish and collect the pollen from one of the trees. We then walk over to another variety (we have three) and lightly paint the pollen onto the green sticky bulbs of flowers still in their female phase. Each tree gets some fresh pollen from another tree several times over the course of the two to three weeks they are in flower.

Our pawpaws will set some fruit without hand-pollination, which we discovered when we saw a few fruits higher in the tree than we are able to reach with our brush. Because of this, we also learned that hand-pollination will result in numerous clusters of four to six (even twelve, as we had one year) fruits instead of the sporadic single or double sets produced by carrion insect pollination.

By October, pawpaws swell to an enormous size. They seem to me as large as mangoes, though realistically a potato is a better comparison. Their green skin takes on a slightly yellow blush and some brown streaking like a ripe banana. When they become a little bit soft and aromatic, we might eat them right off the tree or bring them inside for a day or two to ripen. Different varieties of pawpaw have different flavors. Our seedling tree has orange flesh and a strong flavor like an overripe banana with a bit of mango. This almost raunchy flavor can be reminiscent of the ultratropical durian. Our grafted Shenandoah and Rappahannock trees, developed by a hard-core pawpaw breeder in Pennsylvania named Neil Peterson, have larger fruits with fewer and smaller seeds. Some days I think I prefer the white-fleshed types, which have a milder flavor and a creamy texture and tasted to me like a mix of avocado and pear the first time I tried them. Now I would say they resemble their tropical relative cherimoya. Few visitors to our garden can believe pawpaws are actually native as far north as New York City and Michigan.

Recently we had so many pawpaws we could not eat them all fresh. Meg, Marikler, Jonathan, and I spent a long night cutting them open, removing the seeds, and scooping out the flesh to freeze. (Still, Megan ate so many that she could hardly bear to look at them the following year.) We served them in smoothies a few weeks later at a backyard workshop, and people went nuts for them. Meg also made a pawpaw pudding that year and served it to the Massachusetts Commissioner of Agriculture, who was visiting a farm-business-planning course that she and Jonathan were attending. He was impressed by how tasty the pudding was and could not understand why pawpaws are not in commercial production in our state.

The next serving in our fall banquet of fruits is the hardy kiwi. We haven't actually been able to eat the ones from our own garden yet, though one of them did flower, and we have high hopes of their setting fruit soon. Our friend Steve Breyer at Tripple Brook Farm is generous with his kiwis, though, and somehow we usually end up visiting around this time of year. In fact Jonathan and I do a workshop every fall that teaches forest gardening through feasting on fall delights right off the tree at Tripple Brook and Paradise Lot. Last year our class of twenty people arrived at the peak of kiwi season. Students were amazed at the explosion of flavor in the perfectly ripened kiwis Steve shook down from his trees. Kiwis need to be a little wrinkly and always look to my eyes like they are overripe, but that is when the aroma and flavor is perfect. Just as an alpine strawberry is an order of magnitude more flavorful than ordinary strawberries, hardy kiwifruit taste like fuzzy kiwis on steroids.

Steve has a lot of room and is able to train his kiwi vines on large, established nut trees that turned out to be bad varieties or maples he doesn't need. Jonathan and I cannot allow our kiwis to grow seventy feet tall, so we constructed an arbor and keep them fairly tightly pruned. I have read that under this kind of pruning regime, trellised kiwis can yield one to two pounds per square foot. Given the size of our arbor, that should give us one hundred fifty to two hundred pounds of fruit per year once the vines mature.

The hardy kiwi is not just a novelty fruit but, I would say, one of the world's finest fruits. In the tropics I've eaten fresh rambutan, mango, papaya, mamey sapote, and many others. The hardy kiwi ranks up there with the best, perhaps only slightly below a perfectly tree-ripened mango. Every fall Jonathan and I kick ourselves for not having planted our kiwis first thing, as we knew they take three to five years to bear.

The hardy kiwifruit is a perfect example of the conflict between native plant conservationists and permaculturists. Kiwi vines are from Asia and, there as well as here, climb very aggressively and smother trees (much like some of our native vines, indeed vines around the world). After more than a century in cultivation in the US, at least two small populations of hardy kiwifruit have shown up in the wild. These small instances have alarmed conservationists, and indeed the vines smother everything around them in a frightening way. However, hardy kiwifruit does not disperse great distances or in great numbers like bittersweet, it just makes a scary mess where it does show up.

We need to back up and think about the big picture before we race to condemn hardy kiwifruit. Where does our vitamin C come from now? It mostly comes from Brazilian citrus, and we burn up lots of fossil fuels getting it here, contributing to global warming. Looking at the big context and our need to drastically increase regional self-reliance to address climate change and replace our disastrous food system, hardy kiwifruit makes a lot of sense. In my mind it is as essential a species for cold-climate agriculture as carrots and apples, two non-natives that have naturalized here in the Northeast. If we were to reboot Northeast agriculture, would we not bring apples and carrots to the Americas? I hope we would.

At the same time it is possible to be more responsible about kiwifruit growing. For example, the super-hardy kiwifruit is much less aggressive and has not naturalized. I'm not recommending this species in my workshops as an alternative. There are also some self-fruitful varieties of regular hardy kiwifruit that may have sterile seed. This is rather common in many cultivated species. Certainly breeding a line of sterile-seeded kiwis would not be terribly difficult and could satisfy everyone. That's the conversation I think a rational and reasonable bunch of citizens would have in a society that looked at a wide range of environmental priorities.

Chestnut season always makes me a little sad. Mature Chinese chestnuts are about thirty-five feet across, and two of them would just about fill our entire garden. If we could have swapped out those Norway maple trees that hang over the north side of our garden for anything, it would be chestnut. We did, however, identify some chestnut trees in the neighborhood that we began to harvest from every year. Jonathan and Megan refined a technique for processing them. First they immerse the nuts in water heated to 120°F for about twenty minutes. Then they dry them for two days and put them in the refrigerator. They bring them out and let them season for four to six days before cooking. This allows the chestnuts to dry a little more and makes them both tastier and easier to shell. When it comes time to cook them, we cut an X in the top or a diagonal through the base and bake them at 375°F for about twenty minutes. Once or twice we have forgotten to cut one or two of them and were treated to dramatic explosions in the toaster oven.

A perfectly cooked chestnut it is a must-have experience for anyone who lives where they will grow. Often in the fall we will all sit down and each eat ten or twelve at a time. The aroma is wonderful, and the flesh cooks to golden yellow. Before the arrival of chestnut blight, American chestnuts were the primary canopy species for much of the eastern forest. I cannot imagine what it would be like to have the forest floor littered with chestnuts in the fall.

Since we don't have room for full-sized chestnuts, we grow a shrubby native relative: chinquapin. Chinquapins are small and precocious. We have two that began fruiting when they were about four feet tall. They maxed out at six feet, producing about two hundred nuts from the two little bushes. Unfortunately, the nuts are fairly small, about hazelnut size, and though sweet, they are a bother to shell. Recently I visited the Gold Ridge Experiment Farm in Sebastopol, California, a plant-breeding trial area established by the famous plant breeder Luther Burbank, where many huge trees he bred are on display. There were chestnuts more than one hundred years old, each with multiple branches grafted onto the stock to test out the varieties he developed. One had seven nuts in some of the burs, as opposed to the typical three or four. Seized by inspiration, I hopped on the Internet on my phone right then and there and learned that chestnut can be grafted onto chinquapin. Jonathan and I are planning to try it even if it only results in a small number of chestnuts. The thought of having a bush-sized chinquapin with large chestnuts grafted on to some branches would be a fine feather in our plant geek hats.

By mid- to late October, our American persimmons are in full swing. This species exemplifies the permaculture approach to food production. Ours is growing in some of our worst compacted clay soil, yet it puts on four or even five feet of growth every year. This underutilized native species has received a little breeding attention but should be much more widely grown. For us, it has no pest or disease problems. The fruits are smaller then Asian persimmons, with a more datelike consistency. When underripe, they leave a horrible, chalky, astringent taste in your mouth. But a truly ripe American persimmon, with the texture of an overripe tomato, is a marvel of the culinary world. For me, they combined the best of a perfectly cooked winter squash or apricot and a Medjool date. We eat them off the tree well into November.

Our persimmon tree is about twenty-five feet tall, and with luck it may someday be forty feet or even higher. When I saw a mature American persimmon the size of an oak tree at the Arnold Arboretum in Boston, I was reminded that our forest garden is still young, and we really have yet to see what kind of yields it can provide.

=====

21

A NOURISHING NECTARY NEIGHBORHOOD

You may have noticed that our version of "low maintenance" involves a lot of work on the planning and design end. Nowhere is this more so the case than in preventing and controlling pests. But by 2006 the balanced herbivory ecosystem we designed was more or less functioning.

Birds, shrews, toads, and snakes can offer pest control at the garden scale. Though neighborhood cats take a toll on these animals, we still have an avian and amphibian presence in the garden. Our pond provides water; our thicket-forming shrubs and dense plantings of perennials provide cover. Between our bamboo and the neighboring arborvitae hedge, we have lots of evergreen cover (though to be honest, I don't see nearly as many birds in the bamboo as in the arborvitae). Rotting wood, rock piles, and mulch all serve as additional important habitat and are part of the structural diversity a balanced ecosystem needs. Of course our lack of toxic sprays is an essential component of inviting beneficial organisms to dine.

Not all plants are alike to beneficial insects. Studies have shown that some plant species are particularly desirable as egg-laying, shelter, or perching sites. In our garden we selected borage, comfrey, anise hyssop, and yarrow because they are desirable to arthropods, a key component of our pest-control team. (All have woolly leaves; I would love to see a study comparing woolly and nonwoolly leaves to see if this could be used as a predictor for desirability to beneficial insects.) Lacewings and parasitoid insects prefer the foliage of legume and umbel family plants—both families of which already have a prominent position in our garden for nitrogen fixation (legumes) and food uses (both families). The beneficial insects these plants attract do a great job controlling aphids and caterpillars, though slugs and beetles are still a problem for us.

The most visible aspect of our pest-control program is also the most beautiful. We try to have flowers that provide nectar for beneficial insects constantly blooming in our garden, month after month. As a part of my work on *Edible Forest Gardens*, I compiled a bloom calendar for such flowers including almost two hundred species. I drew this information from a variety of sources that rarely agreed with each other, so I was not surprised to see when we planted these things in our garden that I was going to need to learn more about their seasonal niches in our climate.

Our very first nectary flower in spring is fuki. The fat buds swell under the snow and begin flowering in late March, lasting almost a month. Nothing else happens for quite a few weeks, until mid-April, when a new round of flowers begins. This group includes some of our favorite umbels like cow parsnip, sweet cicely, black Ravenswing wild chervil, and earth chestnut. We also have flowers from our native aster family groundcovers green and gold and Nana dwarf coreopsis.

Early summer brings a heart-stopping display of pest-controlling beauty. Skirret and water celery, mountain mint, and yarrow keep parasitoid and predatory insects supplied with nectar. Udo, a gigantic perennial vegetable from Japan that thrives in shade, forms flower spikes up to eight feet high that are covered with an unbelievable array of wasps. Udo in full flower, buzzing with predatory wasps, is like no flower I have ever seen.

We have a little gap in late summer perennials. The heavy spring emphasis of umbel family species has passed, and the fall power of the aster family has not yet come into full effect. For this reason, we let some Queen Anne's lace go to seed every year. This biennial has a long flowering season through midsummer and helps keep the insects going. Though I have not seen any research on its ability to attract beneficial insects, garlic chive also flowers strongly during this period and is busy with a variety of insects. We used to have a native sweet goldenrod that flowered at this time as well. Unfortunately, this species, also known as blue mountain tea for its wonderful anise-flavored leaves, died on us. It's hard to imagine a wimpy goldenrod, but we just could not get that thing to grow.

When we first arrived at our house in Holyoke, heartleaf aster was thriving in the shade of the Norway maples and along the fence lines. This native species has become part of the urban flora, and we tolerated it for its insect nectary capacity. That still means weeding out about 90 percent of it every year, because it seeds itself around quite excessively. We also had a bit of goldenrod along the fence line, which we left in place, as it is one of the finest beneficial insect plants for autumn. As fall fully kicks in, our Purple Dome New England aster begins to shine. Regular New England asters are tall (up to five feet) and will sprawl and smother plants around them, but Purple Dome is a well-behaved dwarf. We have some planted in with asparagus, and the contrast of the purple aster flowers and silvery asparagus foliage is breathtaking. The last nectary species to bloom in our garden is typically Jerusalem artichoke. Some varieties begin to flower as early as September, but many continue until an extremely hard killing frost knocks them back in November.

All of this is not to say that we don't occasionally lose a crop to insects or that we never do some handpicking. But I take it as a measure of success that we haven't sprayed for insects since about year two (and even then we mostly used just a little hot pepper sauce mixed with dish soap). Sometimes it takes a degree of patience. For example, every spring we have an outbreak of aphids on our beach plums. In time, the ladybug and lacewing larvae show up and eat those aphids and form the populations that will be controlling pests throughout the season. Our first impulse is always to kill the aphids (who likes to see a favorite fruit tree suffer even a little?), but we have learned to observe for a long time before taking any action. Fortunately, this philosophical approach lines up well with our lazy management style.

FOOD FOREST FARM IS BORN
by Jonathan Bates

Like many forest gardeners, Eric and I originally intended to grow lots of food and other useful materials. In time, however, the garden showed us a new kind of bounty, one that may turn out to be more powerful and nourishing than the fruit the garden was initially designed to grow: good livelihoods.

Years earlier, when we were living in Southampton, Eric had planted a seed in my mind: selling plants might be more successful than selling seeds. As the plants in our Holyoke garden multiplied, so did my ideas.

Some people think when plants reproduce in overabundance they become problem weeds in the garden. Could we turn this weedy plant "problem" into a benefit? I was thinking that selling plant propagules (seeds, root cuttings, divisions, etc.) might make our garden system even more productive, bringing both health and wealth into our lives. I had already experimented with selling some of our extra plants for ten dollars apiece to visitors. Might a nursery be more profitable than a seed business? Before I could answer that question, I needed to figure out a few things. I was working full time, had accumulated some personal debt, and was courting Meg. How the heck would that life turn into the one where plants would pay the bills? Could I challenge myself to leave the "rat race," the only work life I knew?

I liked my day job as an energy consultant, but my soul has never flourished working for other people. I remember the day when I was finally able to "let go of the rope," as Meg and I joked with each other. You can imagine the adrenaline that was running through my veins that day. Giving notice was one of the hardest things I've ever done—not so much because I feared what my boss's reaction would be or that I might be burning bridges, but because I was leaving a perfectly good job during an economic recession for a dream that hadn't finished unfolding in my mind.

As a result of a two-year hobby nursery experiment, spending many months rethinking the forty-hour workweek, a little human intuition, and lots of support from friends and family, Food Forest Farm was born in spring 2010.

To make it work, Meg and I needed to learn the nitty-gritty of running a business. I knew I couldn't simply hawk my plants on the street corner and hope to create enough business to feed us. What about accounting, nursery production, customer service, knowing my markets, legal issues, insurance, investments, and business planning?

If there is one thing I would pass on to a budding plant nurseryman, it is this: take a business-planning course, particularly one oriented toward agriculture. The Massachusetts Department of Agriculture Resources (MDAR) offered a program called "Tilling the Soil of Opportunity" that helped me create a plan and guided me toward building a viable business by gaining perspective on the industry and what other farm businesses were succeeding at and struggling with.

The word "live" is in "livelihood." I want real work to be a seamless part of my life: I don't want to get up to go to work; I want to get up to live my life, a flourishing, abundant, and more freeing life. If making money is part of that, then I want to feel good making money that inspires others to create a better world. Like eating, I want my life's work to be fulfilling and nourishing.

In 2011 Meg and I traveled to High Falls, New York, for the seventh annual Northeast Permaculture Convergence. It was a hot summer day when we set out on the two-and-a-half-hour drive. The cabin of the truck reached 102°F, and the plant trailer wasn't much cooler. We needed to reach High Falls quickly and find the nearest shade or the plants would wilt away in the metal oven of the trailer. We did, and the plants made it through the weekend. Folks at the convergence couldn't get enough of the plants and asked tons of great questions. Meg and I found that taking the business directly to our customers was a step toward success.

After evenings swimming in the river, staying up at night gazing at the moon by the fire, and eating fabulous food with wonderful people, Meg and I fell even more in love with each other. We saw friends, taught a workshop, learned new ideas, and sold out of plants. This is the good life. This is the kind of livelihood I wanted to create.

Helping to get more edible forest gardens planted seemed like a beautiful solution to our excessive baby plant problem. Now the nursery expands the wealth to others, which the plants in our garden are providing essentially free to us. We consistently sell out of plants through our website, PermacultureNursery. com. Food Forest Farm was profitable after two years and officially became my livelihood.

22

PATTERNS OF NITROGEN FIXATION

I'm fortunate that in my work I have the opportunity to see many different forest gardens around the world. I find myself observing some metapatterns in how people lay things out. One of the most interesting to me is how people choose to integrate nitrogen-fixing species. Though the species may be dramatically different, I see only a few broad patterns that people use.

One pattern I find fascinating is the nitrogen-fixing overstory: large nitrogen-fixing trees making up the canopy, with productive food plants below and between them. Martin Crawford uses high-pruned alders as a central organizing pattern of his forest garden in England. In one patch at our house, we use a mimosa tree. The first time I saw one of these as a child, with its lacy foliage and pink puffball flowers, I thought it was a truffula tree from Dr. Seuss. While the species has acquired a bad reputation farther south for naturalizing too successfully, here in Massachusetts it is at the northern edge of its range and has never set a seedling in our garden. It creates a light shade that combined with its fertilizing capacity creates an environment underneath conducive to food production. We grow many shade-loving crops beneath our mimosa tree, including currants, jostaberries, edible hostas, giant Solomon's seal, and mayapple. Our bamboo and fuki are also in the shade of the mimosa. Most exciting to us was planting an Arctic Beauty kiwi female, a species much smaller and less vigorous than the larger hardy kiwi and tolerant of partial shade.

Another pattern I often see in edible forest gardens and other agroforestry and food forest systems is to alternate every third or fourth food-producing tree with a nitrogen-fixing tree. In a broad-scale system there is plenty of room to do this, but in our tiny garden we can't afford to sacrifice that kind of space to nonproductive trees. This is why we are so happy with our goumi, which pulls its weight by setting heavy loads of fruit in partial shade while fixing nitrogen.

Like many forest gardeners around the world, we try to put our nitrogen-fixing species to work in other ways as well. We use several as living trellises for productive vines like kiwis and perennial beans, though so far our choice of Siberian pea shrub as a living trellis is far from ideal. Both species have failed to demonstrate the vigorous growth we'd hoped for, though we haven't given up on this model.

More successful has been the "chop-and-drop" system popularized by Australian permaculture guru Geoff Lawton. In this model nitrogen-fixing trees and shrubs are cut to the ground repeatedly in a process known as coppicing. Many species resprout vigorously. The nitrogen-rich leaves are then used as mulch, and cutting back the plants causes the shedding of many nitrogenous nodules that subsequently boost the fertility of neighboring plants. At one point we planted some red alders as a trial. They are killed to the ground by cold here, which we call frost coppicing, but their mature root system allows them to bounce back quickly, with one of ours achieving nine to ten feet of growth in a single season (more than a new-planted seedling could possibly achieve). We have a single plant called Kirilow's indigo with beautiful pink flowers that was initially a companion to our Asian pear, but when the Asian pear began to shade it out, we moved the Kirilow's to the center of a polyculture of perennial ground-cherries to fertilize them.

A new nitrogen-fixer we are excited about is bush clover. Though it has a reputation as a fearsome weed, bush clover is well adapted to our urban soils. The species can resprout seven feet in one season and is adapted to being cut several times a year for mulch, which is ideal for us. Our aim is to cut it frequently enough that it never has the opportunity to set seed and become an unwelcome weed. It demonstrates vigorous growth for us even in virtually full shade. It also has edible leaves, though they are nothing to write home about.

To provide all of your nitrogen in the overstory, you need to dedicate between 25 to 40 percent of plants in full sun as nitrogen-fixers. In the shady understory, that figure doubles, which means an awful lot of plants. Jonathan and I never set out to fix all of our nitrogen, but every bit helps. The problem is that many understory nitrogen-fixers are so aggressive that they don't allow for much else to be going on alongside them.

Two of the nitrogen-fixers we grow in the shade are unfortunately quite aggressive. Hog peanut is one that we have eventually quarantined to certain corners of the garden, where it forms a carpet of beanlike foliage and flowers. Licorice milk vetch has also made itself excessively at home in our understory and now has certain areas reserved for it. In both cases the nitrogen-fixing species will swamp anything less than about two feet tall, so they make a fine understory for shade-loving berries like currants but not a good companion to lower perennial vegetables or gooseberries.

We also played around for a few years with some native shade-loving tick trefoils that we had collected seed for in the woods. They grew well, but the drawback with tick trefoils is that the seedpods stick to you like Velcro and can be almost impossible to remove from a sweater. Perhaps in a larger forest garden they would make a fine understory, but we really can't have such a nuisance in our narrow labyrinth of pathways.

In every part of the world, there is some kind of nitrogen-fixing perennial vine that also produces food. Here in the Northeastern United States, we have several, including groundnuts, which can take partial shade. Groundnuts run aggressively and sprawl all over everything less than six feet tall once they get going. As with hog peanut and licorice milk vetch, we are now designating certain areas of the understory for groundnuts, especially those that do not have chop-and-drop or productive nitrogen-fixing tree and shrub elements. We are also growing a native edible perennial wild bean that like groundnut likes partial shade but luckily stays put rather than running everywhere. This wild bean can grow up to twelve feet high, so we are working on getting some living trellises in place for it. At two years old, ours have not yet reached that size. The beans are small, perhaps lentil sized, and while it might not be worth growing as a food plant on its own, the species has been and can be crossed with other forms of edible beans, offering the potential of a productive, large-seeded perennial edible bean for cold climates.

The last nitrogen-fixing pattern we have trialed is using foot-tolerant groundcovers, such as white and red clover, in pathways. The red clover, which we cut and feed to chickens, came in as seed on some compost. White clover is more problematic. Initially we wanted lots of it because it can take some shade and makes a nice groundcover. Unfortunately, it runs excessively and smothers many of our smaller plants. Our friend Jono Neiger ran an experiment with about ten different species of perennial vegetables and a white clover understory. Those that were tall enough, including Turkish rocket, Welsh onion, and garlic chives, thrived. But the clover swamped other shorter companion species like sorrel, sea kale, and alpine strawberries. A well-behaved alternative to white clover is prostrate birdsfoot trefoil, which grows about one inch high and spreads about two feet across. You can walk on it, and unlike ordinary weedy birdsfoot trefoil, it is sterile and will never set seeds around the garden. The drawback is that it must be vegetatively propagated, a laborious process. Jonathan and I have fantasized about filling our pathways with this species, but the expense and labor have held us back.

Clearly, we have yet to stumble upon the perfect arrangements for nitrogen-fixers in our garden. The fact that the species make their own fertilizer means they tend to grow aggressively. This is as true for native as it is for exotic nitrogen-fixers. One of the key factors in our polyculture design process involved thinking about what nitrogen-fixation strategy to use for each patch. If the polyculture being designed is under the mimosa, we don't need to do anything else to fix nitrogen and can have a fully edible understory. If edible trees and shrubs are the canopy, we need to think about a nitrogen-fixing understory, though that places limitations on what other perennial vegetables can be grown in the shade. Integrating chop-and-drop shrubs seems to be a good compromise in many cases. We can bring in plenty of compost to make up the difference, and even if things don't grow well, we can always walk to the supermarket down the street, so a level of experimental dilettantism is perfectly fine for us. This is one of many cases where we hope our experiments allow others to make newer and more interesting mistakes rather than repeat those that we have already made.

23

GROUNDCOVER CARPETS

It took three or four years for groundcovers to play a significant role in our garden. Jonathan and I were on a budget, so we ordered small amounts of many species of groundcovers to see which ones would thrive. While we were waiting for them to get going (and watching most of them die), we filled in the space among our other perennials with wood chips. But of course Jonathan and I don't like to do any more work than we have to, and loading wood chips into a wheelbarrow and spreading them around definitely counts as unwanted labor. Groundcovers are a natural alternative: they serve as a living mulch, protecting the soil and suppressing the germination of weed seeds.

The ideal groundcover for us is an evergreen (so as to suppress weeds and provide cover year-round), shade tolerant, native, edible, dense, low growing, and able to spread quickly to fill in gaps among other plants without becoming a weed itself. Nothing does exactly all of that, but several species come close.

We have some interesting groundcover candidates that grow to medium height in full sun and also taste good. Sylvetta perennial arugula does not run but self-sows well and is dense enough that nothing else can grow beneath. Chinese artichoke is a mint relative with edible tubers that can produce a dense cover in full sun. Its main drawback is that for best long-term harvest the tubers need to be dug every year, which is often incompatible with the needs of surrounding plants. Perennial ground-cherries, of which we grow a number of native species, can fill in densely in full sun but don't emerge until June. For this reason, we are trialing them with some spring-emerging, shade-loving violets in the hopes of creating an edible groundcover polyculture.

There are many midsized, running groundcovers for part shade. However, these species are so aggressive that it can be difficult to control them. Water celery is a nice edible green in spring and fall, but it can really get around if it is watered or pampered, so we stuck it in a dry place to limit its spread, and we feed a lot of it to our chickens. The many varieties of mint we grow are similar; we don't allow them to mix with our more preferred perennial vegetables and low shrubs because the mint wins every time.

One of our most successful groundcovers for heavier shade is large-flowered comfrey, one of the few plants that thrives in the dry shade under our Norway maples. It is also chicken food, a great soil builder, and preferred egg-laying and overwintering habitat for beneficial insects and spiders. We are playing with several native species for this niche as well. Hay-scented fern is a rare vestige of the landscape that existed on our property before the previous house burned down. We keep it back along the fence but enjoy its delicate look and aroma. The nitrogen-fixing hog peanuts I've mentioned also provide dense cover, though they need to be partnered with a spring-emerging species to provide a full season of cover. Our new contender in this category is Allegheny spurge, a native pachysandra related to the wildly overused Asian species but not as indestructible. It is semievergreen here in Massachusetts, which is desirable for us. We put some in beneath a recent planting of gooseberries and bush clover on the north side of the house along with a number of other new native groundcovers we wanted to trial.

I prefer lower groundcovers. Those I've just listed tend to be twelve to eighteen inches high and sometimes compete with perennial vegetables and low-growing fruit shrubs. One of our favorite low species is the native evergreen green and gold. Also known as golden star, this species forms a low, dense carpet and has yellow flowers that attract beneficial insects. Our favorite companion for it is a native dwarf evergreen coreopsis, with orange-yellow nectary flowers. We have also started experimenting with bearberry, or kinnikinnick, which is native from Cape Cod to California. Horticulturalist Michael Dirr calls it the finest native groundcover of North America. Our plants are small, but in a traffic island at the end of our street there is a planting that has formed a solid evergreen carpet and is thriving in the less than ideal conditions. I've written about prostrate birdsfoot trefoil above under nitrogen-fixers, but we really do love this species for full sun, as nothing else we have found grows only one inch high and spreads to form such a nice, foot-tolerant, nitrogen-fixing mat.

At first thought it seems that strawberries would be an ideal groundcover for us. All of the species are evergreen and low to the ground, and all but alpine strawberries run. The problem has been that at our scale they spread too fast. They also stop producing much fruit for us after a few years, probably in part because they become shaded out as the overstory matures. Nevertheless, in the right area we are happy to have them. Oikos Tree Crops, one of my favorite nurseries, offers many selections of native wild strawberries. Kelly's Blanket is a variety selected for its ridiculously fast rate of spread. It makes a phenomenal groundcover, but we have had to exile it to our fruiting hedge in the front yard because it is so dominant. The tiny fruits are so tender that they break apart when you pick them, but the flavor is wonderful. Also moved to this far corner of the garden is an Oikos selection of the native wood strawberry called Intensity, chosen for its large and flavorful fruits. Like an alpine strawberry, each small fruit really packs in the taste. We hope next year we will be able to sample our musk strawberries. These are picky enough that you need male and female plants, but apparently they taste like pineapples, which sounds worth the trouble to me. We don't use cultivated strawberries as a groundcover because we want to rotate them through the garden in our annual beds to keep disease at bay. We keep our groundcover strawberries far away from our annual beds to prevent them from interfering with our cultivated strawberry rotation as well.

When we were moving to Holyoke from Wonder Bread Farm, Jonathan and I dug up some violets that had been growing in the middle of the grass parking lot there. We knew this species was as tough as nails, though at the time we had no idea what it was. It was only this year when I read *Wild Urban Plants of the Northeast* that I was able to identify them as the native common blue violet. This species has become a major presence in urban areas and is part of the newly emerging hybrid urban ecosystem. We brought a few here and tucked them in to occupy space until we found something we liked better to eat. Common blue violet began seeding itself around our garden and spreading to form mats up to four or five feet across when we weren't looking. This was easier to do than it sounds because it will come up in full sun in spring but can be completely buried under other crops for the rest of the season—yet still continue to spread. It has thick rhizomes and forms a dense mat, and I fear that it may cause excessive competition in the root zone. But we have spread quite a bit of it around the garden because it is such an indestructible groundcover. The leaves and flowers are not as good to eat as Rebecca violets, but they serve in a pinch.

If I were to predict what groundcover would be occupying half of our perennial understory in the coming years, it would be barren strawberry. This native strawberry relative exemplifies every aspect of our ideal groundcover except that it is not edible. It grows low and forms a dense evergreen carpet. It thrives in sun or shade and spreads rapidly without making a nuisance of itself. And it seems to be well adapted to our conditions. I have seen vast swaths of it at botanic gardens and hope to see it do that here in coming years.

It was not until 2008 that we were able to start planting shade-loving groundcovers. The area under our persimmon was the heart of our first experiments. Wild ginger is a native for full shade, but I hold a bit of a grudge against it. I once confused it with another ginger-flavored groundcover and ate the leaves by mistake. Throwing up half an hour later was not my proudest plant geek moment. The rhizomes of wild ginger taste nothing like culinary ginger and are not actually even related to them. Nonetheless it is a nice cover. Another popular native shade groundcover is foamflower. It does not seem to be as happy here as I have seen it in other gardens, perhaps because it does not like our urban fill soils. Nonetheless I remain hopeful that it will form a nice evergreen cover and provide us with beautiful spring flowers in the future.

A rather obscure native species we are playing with is woodland horsetail, a low- growing dwarf of an ancient lineage. In the days of the dinosaurs, treelike horsetails dominated before the advent of flowering plants. Now they are relegated in our garden to grow three inches high and spread in the shade of our lowbush blueberries.

I have been seeing partridgeberries my whole life on walks in coniferous woods. This species exemplifies the mat-forming growth pattern. It reaches only about one inch high and spreads to form a dense evergreen rug. It has an edible fruit, but just barely: insipid, tasteless, and tiny, they are not worth the effort to eat. More interesting from a culinary point of view is wintergreen, or teaberry. Another native, evergreen, shade-tolerant runner from coniferous forests, the species grows higher and spreads faster than partridgeberry, though its berries are similarly tiny.

24

THE GARDEN'S IMPACT BEYOND
THE FENCE LINE

Jonathan and I have been giving the occasional tour to diehard permaculture classes ever since our Wonder Bread days. But around the year when our berries started to look impressive, we began to be flooded with requests for tours. Multiple permaculture design courses, an ecological landscape design school, garden clubs, and student groups, as well as individuals who had read *Edible Forest Gardens,* came to see us.

We didn't design our garden for tours. If we had, we might have included an area for groups to meet, wider paths, and perhaps even space for a tent to keep people out of the rain. Still, Jonathan and I developed a shtick and began offering visitors plants and books for sale afterward. We were also paid for most of our tours. At some point we realized that these paid tours, along with the sales of books and plants, was making a reasonable contribution toward paying the costs of having and maintaining the garden.

One day as I was showing the tropical front yard to a permaculture class, the teacher pulled me aside and said, "We may have a problem." One of the students, perhaps thinking that everything must be edible in an "edible" landscape, had eaten some castor bean leaves. We grew castor beans because they are beautiful and tropicalesque; the seeds are used in making bioplastics. But the beans are extremely poisonous; the ricin gas used in the 1995 Japanese subway attacks was made from castor beans. I rushed down to look in my poison-plant reference books and found that although there was plenty of mention of the poison seeds of castor bean, they said nothing about the leaves. We decided nonetheless to take the student to the emergency room, where doctors pumped his stomach as a precautionary measure. I called the young man's mother, who was understandably upset, to explain the situation and told her we would be ripping out the plants straightaway. I'm glad to say the castor-bean-leaf-eater turned out just fine, and I'm sure he learned the lesson of that day as well as I did.

Now whenever we do a tour, Jonathan and I begin by saying you can eat anything we hand to you, but you can't assume that everything you see is edible—even if you think it is or you think we said it was. Only once did I see people sneak food from the garden: two fruit robbers shared an underripe persimmon, which was a well-deserved unpleasant experience for them both.

Jonathan and I, along with Ethan Roland, were also working on a project that began to bear fruit around this same time. Several years before, with the help of Dave Jacke and Jono Neiger, we had started the Apios Institute for Regenerative Perennial Agriculture. Our first project was to create a wiki of edible forest gardens. Nobody knows better than I do that what you read about plants in books rarely lines up with how they actually perform in your garden. So the first layer of the Apios website is species profiles. The next layer is a place to profile polycultures, with links from each species to the polycultures and gardens that include them. We wanted to post successful and unsuccessful plant combinations alike, so people could learn from them and replicate the good ones.

Because the forest garden model is still new in eastern North America, we wanted a place for people to share and receive basic knowledge accumulated from multiple garden sites. Our hope is that this information can later serve as the basis for proper scientific study, such as of perennial polycultures with commercial production potential. Finally, we wanted a place for people to profile their gardens, show pictures, and tell stories as a way to build a decentralized network of backyard experimenters who are in touch with each other. When the website was finally launched in 2008, almost all of the content was mine, but over time others have added their contributions as well.

To our surprise our little garden and *Edible Forest Gardens* have had an impact beyond the eastern forest region of the United States that we set out to serve. When I joined the board of Nuestras Raíces in the mid-1990s, their harvest festivals were small affairs for community garden and youth members. By the time I joined the staff in 2004, for the start of the farm project, the harvest festivals had been growing for over a decade.

For the 2004 festival, many of our farmers and gardeners were up at three in the morning to start roasting the twelve pigs that would be devoured by two that afternoon. We built a treehouse with a stage in a grove of silver maples and displayed colorful murals of life in Puerto Rico and Holyoke on the backdrop. Trios on the cuatro, gourd, and guitar played old-time Puerto Rican *jíbarro*, music, and there were folkloric dancers, masked *vejigante* dancers on stilts, and demonstrations of *paso fino* horsemanship. Long rows of tents sheltered vendors selling tropical ice cream, fresh produce grown on the farm, and a range of traditional Puerto Rican foods and crafts alongside community organizations performing outreach.

One day just a few weeks after the harvest festival, I received a phone call from a man named Alan Wright. He told me that he lived in Mexico and was a fan of *Edible Forest Gardens*. He had read in my bio on the back cover that I worked for an organization with a Spanish-sounding name and was curious if I spoke Spanish. He asked me if I would like to teach a course in the cloud forest of Mexico, at a permaculture center called Las Cañadas. When I hung up the phone and told Marikler, she said this opportunity was not one I should turn down.

Marikler and I spent months developing a five-day curriculum, learning useful species for high-altitude tropics, and researching the ecology of the Mexican cloud forest. I was in frequent contact with Alan as well as Ricardo Romero, director of Las Cañadas. As part of the process I reread both *Edible Forest Gardens* volumes front to back, taking copious notes, translating large passages of difficult text, and trying to boil it down to the simplest concepts, since I was not terribly confident in my Spanish.

After a long flight and an endless bus ride in the rain, we arrived in Huatusco, Veracruz at what was supposed to be the site. The workers at an unrelated business at the base of the road were on strike, and it was starting to look like we might have to camp out under trees for the night when Alan and Ricardo rolled up in their pickup trucks and took us up to the site, were Ricardo made us quesadillas from fresh farm cheese and homemade tortillas. Everything was going to be all right.

In the morning I woke up and stepped out of our cabin to see a giant snowcapped volcanic peak towering over one of the finest edible landscapes in the world. Ricardo took us on a tour through their site, where I saw one mind-blowing demonstration after another: biointensive vegetables, food forest, rotationally grazed poultry, cattle on alder silvopasture, timber bamboo plantations, coppiced nitrogen-fixing firewood production, and so much more.

The food forest at Las Cañadas was a revelation to me. Macadamias were among the tallest trees. The canopy also included bananas, guavas, cherimoyas, citrus, peaches, and even spices like cinnamon. Many species of nitrogen-fixing trees and shrubs were interspersed. Until recently, the group had been unable to develop the understory because of an intense pasture grass that outcompeted everything else, but after trying many species of livestock, Ricardo discovered that geese were the solution. He fenced in about an acre of the food forest and stocked it with ten geese for one year. The geese ate nothing but grass and clovers, and they suppressed the grass to the point that other herbaceous species like wild ground-cherries and pokeweed began spontaneously growing. At that point Ricardo was able to reduce the geese to two per acre, which was enough to keep the occasional sprout of grass under control while preserving the perennial vegetables and herbs in the understory. It was an absolutely genius solution, reducing human labor and fossil-fuel inputs while making for happy geese. It was as fine an example of permaculture in action as I'd ever seen.

Ricardo is one of the world's great plant geeks, and we talked for hours in Spanish, English, and Latin about useful plants of the world and what cool species might succeed under their conditions. In the years since, the Las Cañadas food forest understory has filled with taro, ginger, perennial hot peppers, native beneficial-insect-attracting wildflowers, sweet potatoes, and shrubs with edible leaves from nopale cactus to chipilín.

At first I couldn't believe that Ricardo, Alan, and his wife, Paula Kline, had found anything of value in *Edible Forest Gardens* to help them in their cloud forest climates. Dave and I had never imagined the book would be of use to anyone outside the eastern United States. But here I was in the cloud forest seeing some of our ideas put into action. In fact, Alan and Paula had taken one of the design illustrations from volume one (a purely hypothetical drawing) and planted it out for real on several acres. When I showed before and after pictures of our garden, I was surprised to see real tropical gardeners and agronomists impressed at the transformation.

I had never taught for more than an hour before in Spanish, so a five-day course was a push, but I've been a different person since I learned that I was capable of that. People came from all over Mexico and were fascinated by the idea of the forest garden. We worked on refinements to the food forest Ricardo had planted and hiked a remnant patch of old-growth cloud forest in search of patterns to emulate. I made many friends for life on that trip.

Walking down the dirt road with Alan and Marikler after the course, I realized that it was time for me to leave my job at Nuestras Raíces and teach, practice, and write about food forests full time. Running the farm project was a fascinating and constant learning experience, and I loved the team I worked with there. Still, after a few years I had begun to feel the itch to do something else. Visiting Las Cañadas was the push I needed to go for it. I began a six-month process of passing on my Nuestras Raíces farm job to the farm site manager, Kevin Andaluz.

I didn't leave the farm for a glamorous life of travelling and teaching (which lost its luster pretty quickly once I'd spent a few nights sleeping on airport floors). I decided instead to dedicate the next chapter of my life to promoting the power of perennial farming systems to fight climate change. Why use expensive geoengineering projects when useful trees make a great carbon impact while restoring degraded lands, providing food and income for people, and creating healthy ecosystems? Having proven to myself that perennial food production worked on the home scale, I was convinced that farm-scale techniques applied at the global level could solve many problems.

In 2009 two dear friends, Arthur Lerner and Emily Kellert, visited me in Holyoke. These days they run FRESH New London, an urban youth farming project in Connecticut. Back in the mid-1990s they had taken me in during a difficult time, when I was writing about permaculture from an urban apartment with no garden. Walking our garden and seeing me surrounded by fruiting trees and shrubs, Arthur remarked that I was like a mushroom: I had spent years building networks underground, and it was only now that the mushroom popped above the surface and showed the visible (and delectable) results of my work. I knew at this point that our garden had truly passed the establishment phase and made it into the productive period.

PART FOUR

REAP

2009–2012

25

EMERGENT PROPERTY

By 2009 our backyard ecosystem was beginning to demonstrate emergent properties—meaning things were happening that were more than the sum of their parts. Catbirds and robins were daily visitors in the garden. Fruits and perennial vegetables were producing incredible yields with virtually no labor input from us. Cycling compost through our chickens made our soil so fertile that the top several inches of the garden were almost pure worm castings. The garden, in other words, had taken on a life of its own. Meg called our study of this aspect of the garden "agrogeekology," which made us agrogeekosystem managers.

You might remember that in 2005 Jonathan and I had marked the pathways of the garden with spent shiitake logs and dead branches left over from trimming our Norway maples. By 2009 those logs had developed the crumbly, spongy textures ideal for water storage and invertebrate habitat. Sometimes my friend Daniel would come by with his eight-year-old son, Moses. All three of us loved rolling over or breaking apart these rotten logs in search of ant nests, termites, and other delectable items to feed to the chickens or the goldfish in the pond, watching as pill bugs, centipedes, and predaceous ground beetles scrambled for cover. A typical log might have eggs of slugs, spiders, or earthworms; some would be threaded through with bright orange fungal hyphae. There's a certain giddy excitement to this kind of discovery that makes me feel as if I'm Moses's age. I also take it as a sign that life, whether beneficial and detrimental to our garden goals, had taken root in the woody apartment complexes we provided for it.

One summer day in 2009, I was working in my basement office, when Jonathan walked in from outside.

"Do you want to see something really amazing?" he said slyly, his hands curled around an object I couldn't see. Jonathan unfolded his hands to reveal a small, bluish salamander.

"Where'd you find that?" I asked, not suspecting that this rare creature might have come from our own yard. As a boy, I had turned over many logs and seen hundreds of red-backed salamanders but never a blue one. Over all the years in our garden at Paradise Lot, I had never seen any kind of salamander.

"Under a log beneath the persimmon tree," Jonathan replied.

How had this rarity turned up in our urban garden? It seemed unlikely that it had migrated across the street from the trash-filled woods along the cliff. Perhaps it was a stowaway in a pot of soil from another nursery or in some compost materials we'd hauled into the garden. Maybe its eggs had been stuck to the leg of a bird. There was no way to know for sure, but how the blue salamander got into our garden didn't really matter anyway. What mattered was what it signified: our emerging backyard agroecosystem was attracting delicate forest organisms to patrol its understory. A woodland amphibian that could never have survived in our yard in 2004 was now at home in the shade and soils we had created.

Our garden also saw the arrival of unwelcome organisms. In the summer of 2010, Jonathan and I noticed that a number of our crops were showing unusual and alarming symptoms. The leaves of our beans, tomatoes, perennial ground-cherries, and even grasses—crops from totally unrelated families—were pale, speckled, and dirty looking, as if they were covered with a sticky powder. Jonathan and I had never seen insect problems at this scale in all our years of gardening together. We raced to our copy of Cranshaw's *Garden Insects of North America* and found our denizen in the section on leaf-sucking pests: spider mites.

These tiny arthropods have a fearsome reputation, for they attack many crops from many different families. And they were coming on fast. Every day more of our plants looked like they were about to give up the ghost. A friend had given us a new strain of perennial ground-cherry that he said had a larger, sweeter fruit than the wild types. It shriveled and died to the ground in just a few days.

Jonathan and I were beginning to panic. This was the greatest test of our Zen pest-control program we'd ever faced. Were we deceiving ourselves by thinking our garden ecosystem was resilient? Was it, in fact, so weak that it was vulnerable to takeover and attack? But while we raced to identify the pest, investigate our least-toxic control options, and decide what to do, our garden was taking steps on its own.

Three days after we first noticed the sad, speckled leaves, Jonathan pounded on the kitchen door while I was eating my morning toast and tahini. His face was lit up with excitement. "Dude! Spider mite destroyers! I read about them in the insect book, and went looking for them, and there they were."

A form of tiny ladybugs, these aptly named predators had been slowly and surreptitiously building their population in our garden. Spider mites themselves are so small you can barely see them with the naked eye. Their predators are larger: they look like tiny black dots.

A number of plants that had turned ghostly pale began to put on fresh green growth, untroubled by spider mites, while affected leaves shriveled up and fell off. By the end of the season, many of our crops looked as though they had never been attacked. Granted, we didn't get to taste those perennial ground-cherries until they resprouted the following year, and some of our greenhouse crops never recovered. But if we had sprayed, we almost certainly would have killed our spider mite destroyers and disrupted the balance that protects our garden.

These days it seems that everywhere we turn the garden is teaching us the power and grace of letting go and allowing nature to take its course. In the spring of 2009, our Barred Rock hen got broody. She sat on her eggs and would not get up. Of course with no rooster, those eggs were unfertilized and would never hatch into chicks. But Jonathan, Meg, Marikler, and I began to brood as well and thought it would be fun to have some baby chicks around. I spoke to Wilberto Colón from Coroyama Farm, one of the farmers who rents land from

Nuestras Raíces, and he gave us a dozen fertile eggs, which we slipped under our broody hen. One morning we saw tiny wet chicks picking their way out of their eggs; within a matter of hours they were adorable fuzz balls.

Watching our hen, I realized why it was so much work for us to raise chicks in the basement. The hen fluffed out her feathers to make a warm environment, just like a heat lamp. Instead of our awkwardly dipping our chicks' beaks in water for the first time, the hen taught them to drink, peck, and scratch. Every task we had labored at, she handled effortlessly, by instinct. It reminded me of the great permaculture pioneer Masanobu Fukuoka's *One Straw Revolution,* where he says the way he got his farm closer to nature was to do less every year until he stripped his production system to the barest bones, letting nature do almost all of the work.

Sometimes Jonathan and I wonder what would happen if we stopped managing the garden altogether and just watched it grow. What would it look like in fifty years? It would likely have some bittersweet and Norway maple seeding in, but I also think that a lot of what we planted would still be there. The mature food forest would likely be composed of suckering persimmon, pawpaw, and bamboo. Currants, gooseberries, jostas, and raspberries would form much of the shrub layer. And the herbaceous layer would be a riot of mint, fuki, groundnuts, hog peanuts, and sweet cicely. Planting something you know will survive and continue to grow and develop when you are gone is the essence of regenerative design and gardening. We have brought into being a living ecosystem that now has its own instincts about where it wants to go.

LIVING IN PARADISE
by Jonathan Bates

It was in bed one night, warm and cozy under the flannel sheets, that Meg and I decided to create a vision statement for our life together. For months, we had been fantasizing about how we wanted to live our lives and where we wanted to be down the road, but we hadn't yet found a way to articulate it.

One of my friends told me about a process for goal setting called holistic management. Originally developed for managing grazing animals on large tracts of land in Africa to combat desertification, it was then adapted and expanded to help farmers and individuals create holistic plans for their own lives. I found an explanation of the process online. After an hour of thinking and talking about it, Meg and I were able to distill our needs, wants, and dreams into a powerful statement: "We have a simple, creative life, full of family and friends, laughter and love, connected to the land and supported by meaningful work, with the time to appreciate it and experience it fully."

That was many years ago. We are now successfully living this vision. Food Forest Farm became one part of the "meaningful work" that Meg and I share together. Even during long nights of accounting and tax preparation, we know that this kind of work is helping us realize our dreams together. It opened up flexibility and freedom in our lives that we had never known or thought possible.

By being in charge of our own time, Meg and I have been able to create new traditions for ourselves, like acquiring, cooking, and eating most of our meals together. When we share ourselves and our minds more fully, in a slow, healthy way, the grind melts away and a new ethos emerges. Our life has become like our garden: free to sway with the breeze and soak up the elements around us instead of motoring along with time.

The business also ebbs and flows and intertwines with our personal life. The experiments in the garden and kitchen that provide us sustenance also inform the business. Customers ask us how to cook the perennial vegetables we sell in the nursery, and we share the recipes we've created in our own kitchen. Meg started a blog to make these concoctions even more available.

It is the food that directly sustains us. In 2010 I decided to keep a kitchen garden log. In it I tried to weigh, in pounds, the amount and types of food I was bringing into the kitchen from the garden. From what I recorded—and it wasn't everything—I estimated that for a household of four adults, over six months, we harvested four hundred pounds of forest garden fruits and vegetables from the perennial portion of our tenth of an acre, in addition to the many annual vegetables from the annual beds, tropical garden, and greenhouse. Much of that forest garden harvest was from partial shade, and many of our plants are still getting established and should produce more in the future. I doubt even biointensive gardening could do better in the shade and poor soils of this part of our garden. Now that the forest garden is established and productive, we can mostly sit back and watch it live to its potential, feeding us for free. Many of our perennials crops yield well with little work on our part.

This abundance has encouraged us to get creative with our food. The late summer fruit harvest, around the time our beach plums ripen, is one of my favorite times in the garden. One year, our tree was so laden with fruit the four of us could not eat it all, and Meg baked a luscious beach plum crisp. Sometimes there is pawpaw fruit by the bushel. We eat it fresh, frozen, share it with friends and visitors, and even turn it into wine. Grapes did especially well one year. Have you ever eaten frozen grapes? They are cool and refreshing on a hot summer day.

Every Sunday Meg and I plan our meals for the week. We plan according to what we have in our pantry, refrigerator, freezer, and garden—and if we have a particular craving, a new recipe to try, a favorite meal we want to revisit, or a potluck we are attending or hosting. We buy supplemental ingredients from the food co-op or local supermarket. (We don't have an avocado tree—yet.) By expanding our food awareness and networking with other farmers and gardeners, we are building connections for the business at the same time. As the garden and business mature, we become wiser and more in tune with the symbiosis of work and life.

There are times when it seems I am living in a dream. I ask myself, "Is this really happening?" What if this paradise *is* lost? You never know what might happen. Meg is a fourth-generation upstate New Yorker. Rural lands full of summer hay bales, endless starry nights, and fireside chats are in her blood. Having more family around is an important goal for us both. Sometimes I fantasize of other blank canvases, imbued with the potential of animal herds, fruitful gardens, and ecological buildings. Eric,

Marikler, Meg, and I have enjoyed this Paradise Lot adventure. Perhaps folks in both city and country and everywhere in between will create similar places. We are surely not the only ones living in paradise.

26

GUIDING SUCCESSION

I 've always been thankful that I was able to take a full semester of ecology while I was in high school. There I learned many of the classic theories of ecology. One of the most fundamental notions is succession, the change and development of ecosystems over time. I was taught the classic model of succession that so many of us have learned. This model begins with a disturbance. Let's say that a land-slide has left bare soil where a forest once grew. The seeds of annual plants ("weeds") that had lain dormant in the soil for decades take advantage of the sunlight to germinate. Other annual seeds arrive, perhaps carried by wind or wildlife. For several seasons, these annual plants grow densely, creating a patch of amaranth, lamb's quarters, and ragweed. But meanwhile perennial plants are joining the mix. Grasses, goldenrods, and asters, initially unnoticed, begin to dominate the meadow by the third or fourth year. But the succession process will not end at meadow. Shrubs like blackberry and sumac seed in or sprout from roots long suppressed by shade. As they grow, they spread by root suckers and start to shade out the meadow species. Sun-loving trees like birches and poplars seed in as well. These young trees and shrubs create a tangled thicket. In time the birch and poplar trees shade out the shrubs and what's left of the meadow. They grow to form the canopy of a young forest. But this is not the end of the story. Birches and poplars need sunlight to germinate, thus a new generation cannot grow in their own shade. But shade-loving species have been moving into the understory. Trees like hemlock and beech can germinate in the shade. Ferns and violets form a carpet under the maturing trees. As the birches and poplars die out of old age, they are replaced by beech and hemlock. We have now reached the "climax" of this ecosystem, its ultimate, mature end state and the goal it was driving toward all along. Classic succession theory holds that any given geographical region will succeed toward a particular, predetermined climax—in this case a beech-hemlock forest.

Unfortunately, this is not how it actually happens. Ecologists have learned that succession is cyclical. It does not have a beginning, middle, and end. There is no point you can call the climax, and it doesn't happen uniformly across a large piece of land. Nor can its outcome be precisely predicted.

Rather, succession arises from a disturbance to a given *patch* of land—maybe a tree is blown down in a storm, creating an opportunity for sun-loving species. Depending on what roots and seeds are present at this opportune moment, including those that arrived just in time, a race for survival begins. The species most suited to the conditions of that particular disturbed patch survive and thrive, forming the basis of a new community. There is no way to predict exactly what kind of plant community will exist fifty years from now in any given area, though we can make general predictions of the species that may be present. Each patch of an ecosystem thus has its own successional trajectory, or potential future, based on what kind of disturbance happened, what conditions are present in the patch, what species were there when the opportunity arose, and how well different species thrived.

This nonlinear model of succession describes a mosaic of patches, each undergoing variations on a theme with slightly different species present. Each patch may be in a different stage of maturity depending on how recently a serious disturbance occurred. It's a much more fluid and complex model of succession than I learned in high school.

The "middle" stage of succession, which might be a patchwork of goldenrod meadow, shrubby thickets, and scattered stands of young trees, is the most productive time in an ecosystem's life. During this stage established perennial root systems provide the greatest rate of biomass growth of any stage in the eastern forest region. Midsuccession is also the favorite niche of most of our best food-producing perennials. For example, tree fruits from pears to persimmons, shrubs like hazelnuts and raspberries, and perennial vegetables like groundnuts and asparagus are found in midsuccession habitats in the wild. There is enough sunlight coming in between the scattered trees and shrubs to allow for a productive understory, but the diversity of sun and shade allows for lots of different interesting things to grow.

How do you maintain a forest garden in midsuccession? I couldn't imagine cutting down a productive, tasty fruit tree that had finally grown to full size just because I wanted to mimic an ecological phenomenon. Spacing your trees far enough apart so that when they grow up sunlight can still reach the plants beneath them was part of the strategy. Beyond that, I wasn't really sure how this could be addressed.

Having lived with the forest garden for eight years now, I've learned that nature takes care of this theoretical problem for us in the most practical of ways. First of all, just like in a "real" ecosystem, some of our trees died. One of our beds was particularly ill-fated. Jonathan and I knew from the outset that some of the most popular fruits in our region are difficult to grow without a lot of attention to pests and diseases. This is why we chose pawpaw and persimmon (native and pest resistant) to be our tallest fruits and selected mini-dwarf and bush versions of apple, peach, and cherry. We didn't want to waste space on things that might not work out, but we did want to try our luck with these tasty fruits.

Over the three years that it set fruit, our mini-dwarf apple ripened precisely one. Multiple pests and diseases attacked the fruits, and squirrels stole the remainder when they were still sour and green and rock hard. Our bush cherries were delicious but wiped out by stem borers within a few years. Our patio peach set heavy loads of fruit but was devastated by plum curculio. The little scars from curculio wounds on the fruit oozed a thick jelly, as did the multiple stem wounds from borers. Almost this entire bed was a failure.

To be honest, we were chomping at the bit to try some new plants anyway. So rather than adopt an intensive organic spray program, we decided to replant with more resistant species. We dug everything out of the bed and took a stab at designing a new polyculture around some hazelnuts and a nitrogen-fixer with edible fruit called seaberry, or sea buckthorn, a new crop to the Americas, increasingly grown in the northern plains of Saskatchewan and Manitoba. The highly nutritious berries are used to make an orange-juice-like drink consumed through much of the coldest regions of the former Soviet Union. It can handle intense droughts, is hardy to zone 2, and is used for reclaiming strip mines in Europe and Asia. The nearby state of Connecticut preemptively banned it, fearing its potential as an invasive species. We had high hopes that seaberry would thrive in our compacted clay. Wrong. One died right away, and two years later the other had still not grown an inch. The hazels also stalled and refused to grow.

Meanwhile we planted out the understory with things we knew liked sun, such as green and gold and dwarf birdsfoot trefoil. We also planted out a number of tiny seedlings of a new crop we were trying, sylvetta arugula. The arugula exploded into growth, and by the following season it was an unbelievably bad weed, with a carpet of seedlings coming up in the middle of every other plant in the polyculture. It was time for the entire bed to be reinvented once again.

This time we would create a "designed disturbance" that we hoped would help our desired species thrive. We had already seen what a broadfork could do, so we dug out the surviving plants, thinned out the sea of arugula from around them, and transplanted them to their new homes in next-generation polycultures. With some aluminum flashing we created three rhizome barrier beds, which we used to trial our sunchoke polycultures. This year those same beds will become the home of blackberry and black raspberry cultivars intended to fill a two-week gap in our fruit season.

But natural selection was not the only editor at work in our garden. Jonathan and I had wanted for years to grow sandraberry. This vine produces a red berry said to embody all five types of flavor recognized in Chinese medicine. When the fruits were finally ripe, Jonathan and I went out to taste them together. The small fruit was filled with a large pit and reminded me of a bitter cranberry. Jonathan spit his out, too. So did 95 percent of the other people we asked to try it. On top of tasting bad, the plant was a ferocious weed, suckering in every direction and climbing madly on neighboring plants. We ripped it out and replaced it with an akebia vine, a fruit all four of us had enjoyed at Tripple Brook Farm. Another case of "unnatural selection" was the broadleaf culinary sage that used to grow in a prime sunny spot to the south of our persimmon. It grew extremely well and dominated all of the other plants around it, but between the four of us we used perhaps one small sprig per year. I wanted that space to grow more of something we really liked to eat: Turkish rocket broccolis. The other three people in the house claimed they would continue to eat some sage, so we transplanted it to our recycling bin storage area, where it would get beat up plenty. To no one's surprise it has continued to grow there, though at a rate more appropriate for the amount we eat.

So my worries about how to maintain a midsuccessional state in our garden have proved largely unfounded. Between the natural death of plants and our own ruthless selection, every year a few patches start a new successional trajectory. Our garden is a shifting mosaic of patches in different stages of succession, just like a real ecosystem.

IMAGINE SELF-RENEWING ABUNDANCE
by Jonathan Bates

I enjoy a warm summer evening at the local megaplex, experiencing a movie on the big screen with friends or family. The whole adventure reminds me of the good times of summers past. Yet it wasn't the movie that most stirred my emotions one evening. On the way from the theater, through an urbanized and asphalt-filled strip mall, our car passed three large, old, unkempt apple trees. Illuminated by the setting sun, they were overloaded with beautiful, blemish-free apples. When I pulled the first apple off the tree and bit into it, I tasted an almost perfect Golden Delicious. It could have come from the farmers' market. And it looked as if all the apples at arm's length had already disappeared (they must have been very convenient for the firefighters in the station right next to them).

Continuing home in the car, I wondered about those remnants of an old apple orchard. They were out there like little islands, surrounded by asphalt and concrete, cars and rushing mall shoppers. The trees were ghosts from a forgotten time and definitely out of place. Or were they?

I imagined what life must be like for those trees. They looked pretty good for not being managed. All those large, spotless apples—how did that happen? Not much ecology around for apple-pest predators or apple pests to overwinter. The heat of the area must sustain a nice climate for ripening. The trees provide habitat for birds and cool the ground with their foliage. Why not always have abundant fruit trees where people are?

The abundance in our garden comes to us in a similar self-renewing way. Our fruit trees are surrounded not by grass and asphalt but by other useful and edible easy to care for plants. After eight years, with very little care from us, all the plants are providing food, medicine, mulch, fodder, beauty, habitat, knowledge, seeds, and baby plants.

It is the plants that ultimately taught me to see abundance. Like how one tomato seed becomes thousands, or how apple trees can swell with hundreds of apples, with no care, in the middle of a city. How is it that the abundance that I am now seeing in the garden, and in life, was hidden from me all this time? Along with our garden, the neighborhood has become a place for experiencing life's abundance too.

Early on, Eric and I imagined hardy banana plants in front of the house, sheltered from the westerly winds, and collecting the heat from the asphalt driveway. This was their spot because we knew their roots needed to be protected from cold frosty winters. Even without producing fruit, they had the potential to be so fantastic. The leaves are used for many things: wrapping tamales, adding flavor to cooked rice and other dishes, and used as compostable plates.

Now that we've seen many years of wondrous beauty, the bananas have showed us their true nature, as show stoppers! I've watched drivers stop in the middle of the street in front of our house many times. They've become a great landmark for people who are coming to visit as well.

I remember one fall when a Puerto Rican neighbor came walking down the street with his machete to ask if he could cut down the leaves and take them home. It was the evening before the first hard frost, and he was going to take them to wrap pasteles (like a Puerto Rican tamale, stuffed with pork, winter squash, and other savory delights). Lifting the blade high into the air, he swung down, slicing through each leaf stalk with ease. Twenty or so leaves fell to the ground that day. Some pasteles came back in return, a very nice exchange.

Sometimes we can't eat the eggs from our chickens fast enough, so they become gifts for the neighbors. Eyes light up when a dozen eggs, filled with rich, golden yolks, is passed over the fence to them— abundance turns to joy. Who could have imagined!

The sharing of food with neighbors has become a theme in the evolution of this Paradise Lot. Even though we've birthed a successful nursery business, whole plants are also sometimes given away, symbols of self-renewing abundance for friends, family, and neighbors alike.

On the south side of the lot our neighbors have their own garden and lawn. Two grape vines meander along a century old, crusty wrought iron fence, transecting the two yards. One day, while I was observing the vines, I noticed new trees planted in the neighbors lawn. At first it was inspiring to see more trees join the neighborhood . . . and then it hit me, "These trees are going to get big, and they are on the south side of the garden, in five years our greenhouse will be in full shade . . . Yikes!"

Eric's brain came to the rescue though, "Why don't we offer to plant Asian pear trees in place of the maples? That way our garden won't be shaded, they will get beautiful flowers and fruit, and we will have more pollen for our pear tree."

It was a simple, yet brilliant, win–win for every-body on pear tree planting day. The aroma of newly cut grass filled the warm breezes that drifted by. A couple of shovel scoops here, a couple more there, hands plunging into the cool, moist earth, and the maples were out. Soon after that, the pears were in their new home. Just in time too, for the sky darkened, the wind picked up, and giant rain drops started to fall. Claps of thunder finally chased us inside, where icy glasses of lemonade were poured for all.

Over time, it is conceivable that the reality of abundance in our garden, slowly expanding to per-meate my mind, will one day reach out to all the minds in the neighborhood. What would life be like here, if this were to happen? How would the landscape change? Will it be possible to harness the self-renewing properties of this Paradise Lot and expend them outward into the community, city, region, world? Imagine the possibilities.

27

INDIGENOUS MANAGEMENT INSPIRATION

When Europeans first arrived in Massachusetts, they marveled at the large, widely spaced trees with open ground beneath them, through which they raced their horses at high speeds. There was such an abundance of game, nut trees, and berries that many thought God had created a paradise for them. In reality, this open nut forest was the result of frequent light burns by native people as part of a deliberate strategy to maintain an edible ecosystem with little work. In our part of the country, such controlled burns favor the growth of widely spaced chestnuts, hickories, walnuts, and edible-acorn oaks. Blueberries and raspberries resprout vigorously after a fire, benefiting from the removal of disease and pest organisms. Grasses and legumes make pasture for deer, while fire-loving morel mushrooms fruit heavily. Following a fire, human food from plants increases by as much as double, and the amount of game can quadruple.

Unfortunately, the historical record from eastern North America is fairly weak, as most native people were removed from the land, many by disease and war. To my knowledge these cultural traditions of land management are no longer practiced on a large scale by the native people living here today; however, farther west, Wisconsin's Menominee forestry, and the wild rice cultivation practiced across the lakes region, represent the continuation of thousands of years of tradition.

Several years ago I read M. Kat Anderson's *Tending the Wild: Native American Knowledge and the Management of California's Natural Resources*. This five-hundred-page tome is the result of Anderson's years of interviews of native people in California. Many recalled a time when their grandparents maintained "natural" landscapes with active management, and many continue to manage land and useful plant species today and are even having some success in their fight to gain control over "wild" lands they historically managed. While Dave and I were struggling to wrap our heads around the idea of gardening like ecosystem managers, Anderson's book revealed that California's indigenous people had done so on a vast scale for twelve thousand years.

In California (and at the very least the prairies and everywhere oaks grow in the United States) native people used frequent, low-intensity burns to maintain a productive landscape with diverse types of habitats. Fire also killed many pests and diseases of favored plants and was highly effective as rejuvenative pruning for many species. In some cases annual crops like amaranths and edible-seeded wild grasses were sown and raked in after burning. Sadly, the policies of fire suppression and removal of native people from much of California resulted in a homogenization of ecosystems and the near disappearance of certain habitats.

Native Californians applied to "wild" plants many practices I was trained to think of as horticulture. By pruning, controlling pests, weeding, and even irrigating, native people encouraged the plants they wanted to grow to be as productive as possible. Superior varieties of many useful species were transported within and sometimes far beyond their native range and planted in areas where they were likely to thrive. Outside of California, native people took many wild plants into cultivation and even domesticated a number of them. We have native people of the Midwest to thank for the sunflower, one of the world's most important oilseed plants.

What really blows me away is that native people learned to refine the timing and type of harvest so that by digging or picking their food and fiber plants they were ensured a sustained or improved harvest in coming years. Examples range from weeding and replanting propagules of root crops while harvesting to picking cactus pads in such a way that two pads grew back where one had been picked. These management techniques represent a more subtle kind of disturbance.

Rather than knock a plant community back to "restart," they more
subtly shape the character of ecosystems, often maintaining a midsuccession state indefinitely.

If there had ever been any question in my mind that Dave and
I had invented forest gardens or that permaculture was a recent
phenomenon, *Tending the Wild* laid it fully to rest. Native management of the North American landscape may be the largest example
of permaculture the world has ever seen, though there is no reason
not to assume that, as we know was the case in Australia, indigenous
people managed all of the world's landscapes in such a fashion until
the arrival of agriculture.

I've been forced to reevaluate my ideas of "nature," "agriculture,"
and "wilderness." It's all up in the air as far as I'm concerned right
now, which leaves me in an interesting place as a member of a permaculture movement that aims to mimic nature in our gardens but
leave wilderness intact and pristine. The opportunities I have had
to explore indigenous management practices at the Woodbine Ecology Center in Colorado and Occidental Arts and Ecology Center in
California have expanded my horizons enormously, and I'm much
looking forward to learning more in coming years.

As a budding ecologist in the 1970s and 1980s, I learned that the
best we can possibly do as environmentalists is to minimize our
impact on nature. The ideal footprint would be no footprint at all.
That doesn't really give us a lot of room to breathe, and with that as
its model, it's easy to see why the environmental movement has not
won wider acceptance. The most profound thing I have learned from
indigenous land management traditions is that human impact can be
positive—even necessary—for the environment. Indeed it seems to
me that the goal of an environmental community should be not to
reduce our impact on the landscape but to maximize our impact and
make it a positive one, or at the very least to optimize our effect on the
landscape and acknowledge that we can have a positive role to play.

I began to see that Jonathan and I had stumbled across many indigenous management techniques in our efforts to care for and learn from

our backyard agroecosystem. I started to think about how we might more apply some of these ancient practices more intentionally in our backyard. Frequent burns were definitely out of the question, but we can and do use disturbance to keep all the patches in our garden from being uniform. Our tools are the broadfork, sheet mulch, pruning, and tillage. We have taken many native species from our region into cultivation and are part of broader efforts to domesticate some. Certainly we get an A for transporting species beyond their native regions, and I enjoy thinking of this as a practice with thousands of years of history on this continent.

It's around the area of regenerative harvests—harvests that help maintain or improve the productivity of crops in future seasons—that I have most been trying to wrap my brain. Jonathan and I had already learned that some of our perennial root crops need to be harvested every year. This annual thinning reduces competition among sunchoke plants, for example, which actually stop forming tubers if left to their own devices for long enough. Our efforts to implement a coppiced leaf crop system will be another reflection of regenerative harvest.

At this point most of our weeds are the offspring of species we planted here on purpose. In some cases this can make the acts of weeding and harvesting become one and the same. Sweet cicely seeded itself more than we wanted in several areas of our garden. When I learned that the roots are edible, I ran right out and dug up a year-old seedling. At the base was a taproot like a midsized white carrot. When Jonathan and I cooked some up, we were thrilled to discover that they tasted like sweet licorice. Suddenly the annoying chore of weeding sweet cicely was a lot more fun, to the point that by now they are getting pretty hard to find.

Groundnut is another interesting example. It runs and spreads underground throughout the garden and makes kind of a nuisance of itself, climbing on small trees and shrubs. But when I sit back and think about it, the foliage can be cut and laid down as a nitrogen-rich mulch, and the tubers are always ripe and can be eaten any time of the year. I may not be happy to find groundnut in any particular bed, but knowing that I can mash some up like refried beans for a snack makes it hard to be too upset.

My fantasy is that in decades or centuries to come, almost every interaction with one's garden could be a form of harvest, guiding succession gently to an evermore productive future. Some farmers I know have taken this to a markedly elegant level. Steve Breyer of Tripple Brook Farm, my plant mentor, envisions a landscape of fruit and nut trees, with an understory of berries, perennial vegetables, native wildflowers, and dense, long-lived groundcovers. One of the obstacles he quickly encountered in his efforts to implement this strategy was that squirrels were taking his nuts. Steve introduced me to the idea of employing squirrels as a labor force. He has a big old butternut tree with a hollow trunk. Squirrels pick out the best nuts (no hollow or worm-infested specimens allowed) and stuff them into the hollow trunk of this tree for storage. Steve often collects nuts there and introduced me to the idea of trading nuts for corn. This is an idea with indigenous roots that may go back thousands of years as well, with hog peanut and sunchoke recorded as being among the species so traded. I recently met a northwestern hazelnut grower named Rick Valley who does just this. One day he had left out some five-gallon buckets full of sawdust near his hazelnut stand and discovered the squirrels had filled them with nuts. Now he sets out many sawdust buckets every year and allows the squirrels (usually the dreaded enemy of hazel farmers) to select the finest nuts and harvest and store them for him. He provides corn for their wages.

For me, this kind of practice is a tiny window into the complex and diverse edible landscapes that could surround our homes. Jonathan and I already have some beneficial interactions with neighborhood wildlife. Birds eat some of our fruit, but they eat insects, too. Opossums eat rotten fruit drops that would otherwise harbor pests and diseases. Even our squirrels, which are at times annoying, eat thousands of Norway maple seedlings every year. Without their efforts, Jonathan and I would be so stiff we could barely stand from the effort of weeding all those trees. Perhaps someday we'll find a way to harness the digging behavior of skunks or even "farm" Japanese beetles and convert more garden enemies into partners.

Learning about indigenous management and working my own garden shattered the permaculture ideal of the self-maintaining forest garden. Mollison's *Permaculture: A Designers' Manual* offers the beautiful dream of establishing gardens that grow on their own to the point where they don't need us anymore: our role becomes simply to pick fruit and lie in a hammock. I'm all for hammocks and fruit, but I'm learning to embrace the idea of gardens that need us not to toil against weeds and bugs but rather as part of the ecosystem, to hold the rudder and help steer nature in a direction of delightful abundance and elegant complexity.

28

NEXT-GENERATION POLYCULTURES

One spring day in 2009, I gave a garden tour to a young man from New York City who had a forest garden in his tiny front yard. In a ten-by-ten foot patch, he had planted an Asian persimmon and a full set of companions for nitrogen fixation, groundcover, and additional perennial foods. As we walked the garden, I pointed out many species, and we sampled some fruits and greens that he had never tasted before. Usually by the end of a tour, people say how impressed they are with our garden. But this young man had obviously read my books.

"So where are the polycultures?" he asked.

"Um, all around us?" I replied sheepishly. I pointed out a few areas where species were actually playing nicely together.

But he was right. It was our herbaceous layer that was a disaster. When we had put in this garden, we knew little about most of the hundred and fifty or so perennials we were planting. We had a general sense of whether they wanted sun or shade and whether in theory they stayed put in a clump or ran in every direction, but we had little in the way of firsthand knowledge of their mature form and behavior. Nor did we understand the principles that make for successful plant combinations. As a result we just planted them out more or less randomly within the herbaceous layer and hoped for the best.

Within a few years our perennial plantings were a riotous tangle. Plants sprawled over each other. Robust species smothered delicate rarities. Vines like groundnut climbed high into our pear and beach plum trees, competing for sunlight.

Our visitor from New York inadvertently made a point that Jonathan and I needed to pay attention to: we resolved that starting the following year we would remake our polycultures in line with what we had been learning in our own garden and beyond.

Back in 2004 Jonathan and I set some pretty lofty goals. We hoped to create a "backyard foraging paradise," a "megadiverse living ark of useful and multifunctional plants from our own bioregion and around the world." By 2009 we had essentially realized these goals. Our grand experiment had given us a garden with such diversity that we had seventy species of perennials with edible leaves. We also learned that only some of them were really worth eating, and others just didn't want to grow for us. We had cast a wide net and now needed to pluck out the best of what our exploration of diversity had offered us.

We had aimed for three hundred species on our tenth of an acre. At this point we have roughly two hundred perennial or self-sowing species, with perhaps one hundred sixty or so of those edible in one fashion or another. I'd bet that we've tried and killed pretty close to one hundred others over the years. Having met our initial goal of testing maximum diversity, we had to refine things.

We revised our goals to a much simpler statement: "to grow the things we like to eat, that grow well for us, and assemble them in functioning polycultures." No need for bad-tasting crops that grow well or things we love that always die. No need for a tangled mishmash of crops that can't be easily harvested and shade each other out.

The garden itself suggests many polycultures to us. Sometimes this is a case of "Your peanut butter is in my chocolate." For example, we were growing walking onions in one area and Hidcote Blue dwarf running comfrey in another. They grew into each other's areas under a jostaberry and got along wonderfully. After observing for a while, we weeded out a few garlic chives and sweet cicelies in the following spring, pruned up the josta to let in more light, and transplanted a bit of the comfrey and walking onions around to spread the polyculture even further. Two years later both humans and plants seem quite happy with that arrangement.

A lot of forest garden species are a double-edged sword. The problem is that "low maintenance" often translates into more success than you might like. Hog peanut is a great example. This native wild bean can grow in shade and fixes nitrogen, making small edible beans above ground and larger ones underground. A self-seeding annual, it grows from these underground seeds every year. It germinates rather late, then really takes off through the summer. Once it has been in place for a season or so, hundreds of seedlings will come up to form a dense carpet in spring; their viney growth will smother almost anything less than three feet tall, so you need to weed them constantly or cut them back from other smaller plants. We have learned to match hog peanuts with taller species. Under a five-foot jostaberry, hog peanuts are welcome as a shade-loving, nitrogen-fixing groundcover. Under a two- to three-foot gooseberry, they provide excessive competition (and just try weeding them out of a super-spiny gooseberry).

We also learned that some of our species, like ramps and toothwort, or native wasabi, come up early in spring but disappear by June or so. We realized that hog peanuts, which must germinate from buried seed each year, are getting large just at the time that ramps and toothwort are dying back. We already had a nice patch of ramps growing in the shade of our pawpaw trees, and in 2010 I planted some hog peanuts there with good results. I would not have necessarily known that from reading a book; it took repeated observation to figure out the cycles of these plants in our garden.

Harvesting hog peanuts adds another layer of challenge. The below-ground beans are nice for fresh eating or cooking. They're also only about a quarter inch round and covered with a brown fuzzy coating, so they are almost indistinguishable from clods of dirt. (Peeling back the fuzz reveals a beautiful white-and-purple-striped bean inside.) Although books on wild edibles tend to say that hog peanuts grow in the soil, in our garden we find them mostly right between the mulch and soil layers. Digging around to find them messes up the mulch, which is often incompatible with other polyculture companions.

In 2011 we decided to try some new variations on our three-brothers polyculture. In April we removed everything from an unsatisfying bed and went over it with the broadfork. We tried three versions of sunchoke polycultures in beds that had been prepared with aluminum-flashing rhizome barriers. The first was our classic three-brothers combination. The next was sunchoke and woolly bean, a native annual wild bean. The third was sunchoke and hog peanut.

This last was the one that I was most excited about and secretly betting on. Sunchokes meet all of the criteria as a companion for hog peanut: they emerge early and are sizable by the time hog peanuts germinate, and they reach up to eight feet. No chance that hog peanuts could smother them! Hog peanuts can fill in for both the groundnuts and Chinese artichoke, as they both climb and sprawl along the ground as they fix nitrogen. The sunchokes have to be dug every year anyway, so while you were digging them you'd surely find many hog peanuts as a bonus harvest, even though they are not necessarily worth harvesting as a crop on their own. At least the harvest practices of each species are compatible.

My visits to Las Cañadas and other tropical permaculture sites provided some models I wanted to play with at home. One simple pattern that I should have thought of myself is to grow edible vines on nitrogen-fixing trees. There's no reason to waste all that space on fertility when it can also serve as a living trellis.

One of the first things we planted when we moved into the house was a Chinese yam on our front porch. It grew as an ornamental for years and made tiny aerial tubers that I thought of as a novelty crop. After three or four years, I noticed that it was making a lot of those little tubers. Marikler wanted to try them, so we cooked some up, figuring they would just be okay. To my surprise, with a bit of butter and salt they tasted like new potatoes. Today we call them "yamberries," though of course they are not really berries, just tiny tubers growing on a vine above ground. I estimated that our single mature vine produced three or four gallons of little tubers, but the way it was growing on our front porch made harvest difficult.

Jonathan and I talked it over and decided that Chinese yam would be the first vine we would try to grow in a living-trellis system. We had recently cleared out our Regent juneberries and had an empty space in that patch. We planted out three Siberian pea shrubs with the goal that we would prune them to a T or Y shape to serve as a nitrogen-fixing, living trellis, perhaps with bamboo poles laid across them. Unfortunately, even though I knew better, I did something stupid: I planted the Chinese yams at the base of the pea shrubs in the same year, without giving the shrubs a chance to grow and get established before having to compete against aggressive sprawling vines. I've spent the last few years pruning back the vines to try to let the shrubs get established. It also seems as if Siberian pea shrub has persistently failed to take off in our garden. Despite these multiple challenges, I do feel that this model can eventually work. We're going to try swapping out some of the pea shrubs for red alders.

Also in my travels I have seen the benefit of growing something edible and sun-loving in the understory while waiting for trees and shrubs to establish. In 2009 and again in 2010, I had the chance to visit the demonstration farm of the Educational Concerns for Hunger Organization (ECHO) in Fort Myers, Florida. I was captivated on both visits by a living-trellis system that they were establishing using a widely practiced tropical living-trellis polyculture. The long-term design is for passionfruit vines on a nitrogen-fixing, living trellis of madre de cacao, with an overstory of coconuts. The first time I saw this system, the trellis trees were young and the understory was planted out to eggplants. Why not get a yield while waiting? They hadn't even planted the passionfruits yet, nor had they been planted when I returned a year later. Even in the tropics, it takes time to get a trellis tree strong enough to support fast-growing, vigorous vines.

I have also observed this establishment pattern at the Central Rocky Mountain Permaculture Institute in Colorado and the New Forest Institute in Maine. Growing annual vegetables, strawberries, or other crops gives you something to eat while you wait, and sometimes the care like watering, weeding, and fertilizing that you give

those annuals is better than what you might give a somewhat forgotten, not-yet-fruiting tree. Our first priority for our yamberry trellis
understory had to be a nitrogen-fixing groundcover that didn't mind
having a tarp thrown on it so we could shake down the yamberries.
That did not preclude growing something between now and yamberry
maturity, nor did it preclude growing some spring ephemeral species
that will be dormant by yamberry harvest time. For a legume, we
went with birdsfoot trefoil, an urban-adapted nitrogen-fixer similar
to clover. We planted out the area densely with Jonathan's family
heirloom elephant garlic and had nice yields, as it was still essentially
full sun. We also grew some perennial Kurrat leeks, which had been
languishing in the shade for a few years and had never flowered. We
added ramps and edible camas bulbs so that in the long term we can
have a spring yield as well.

Though they look good on paper, plenty of the polycultures we
have tried have not worked well on the ground. We designed an
all-native polyculture to grow in the part shade under our Norway
maples. The goal was to provide a living trellis and useful groundcover support species for a native perennial wild bean that grows in
part shade in tangled thickets. For the living trellis, we selected false
indigo, a native shrub that is a common urban weed in Holyoke that
we expected to thrive. Sadly, after a year of growth it could charitably
be described as six inches tall and half dead. The common blue violets
that we knew were already in our understory thrived here, but our
attempt to grow native wild garlic as the violets' understory companion was a complete disaster. None of the plants seem to have come up
at all, though it has a reputation as a fairly ferocious weed. The beans
did fine, which gives us a two-out-of-four success rate, not our most
shining polyculture demonstration.

One of the best of our next-generation polycultures is the reworked
understory for our beach plum. We love these newly domesticated
native fruits and they grow next to the main path that we use on
tours, so we wanted to show the best we had to offer in this area. We
didn't feel the need to add nitrogen-fixers, as the yamberry–pea shrub

polyculture is right next to the beach plums. We planted a living
barrier of dense Profusion sorrel between our beach plum area and the
yamberry zone, so that the different groundcover systems would not
contaminate each other.

By dividing our green and gold and dwarf coreopsis, we were able
to thoroughly plant out our beach plum understory with clumps about
twelve inches apart. By the end of one season, they were filling in;
it is clear that next year they will form a dense evergreen carpet and
a lovely groundcover. Of course we wanted to see what other food
plants might be compatible with this groundcover. After some thought
we planted bulbs of ramps and camas, as we had in the yamberry
area. Our beach plum zone is showing every sign of shaping up to be
a nice patch of low-maintenance, food-producing, pest-controlling,
edible forest garden. It will also be an attractive corner, with white-
flowering plums, yellow-flowering groundcovers, and blue-flowering
camas in springtime. Every species in this polyculture is an eastern
native with the exception of West Coast camas, which perhaps could
be swapped out for the native eastern variety.

We have a question that helps us evaluate new polycultures on
paper before we go to the trouble of planting them: Is every species
going to be happy with the niche we are providing it? Of course
ideal conditions yield the best production, but sometimes you just
can't create these environments and have to accept reduced yields or
imperfect health. Other times you actually want to slow something
down by keeping it unhappy, an idea I learned from Steve Breyer of
Tripple Brook Farm. We do this with fuki and water celery; it's more
fun to design a way to slow them down with dry shade than to weed
them continually in an ideal, wetter spot.

We are as glad to borrow ideas from other gardeners as we are to
imitate what we find in nature. While replicating polycultures we
have seen in Florida and Mexico involves substituting hardy for tropical species, when we look to Martin Crawford's remarkable English
forest garden we can often directly replicate a species combination.
One that I saw covering a good-sized area when I was there in 1997

is comfrey with mint. This exemplifies the pattern of tall clumping species (comfrey) emerging like islands out of the sea of lower runners (mint). The mint is aggressive and smothers anything smaller than it, but it doesn't stand a chance against a large clump of comfrey. Both plants provide nectar and habitat for pest control, and combined they make a satisfactory groundcover. In Martin's two-acre garden, this polyculture takes up lots of space in the understory. In our case we have little room to work with and are concerned about allowing mint out into the more civilized areas of the garden. Our solution? We planted mint and comfrey at the base of the house along the vehicle access alleyway to the south. The shady conditions help keep both plants in check and being driven over by a pickup truck every once in a while reins in their ambition. We cut the comfrey for mulch and chicken food and use the mints for tea. This polyculture has been growing there for eight years and has remained relatively well behaved and not spread much beyond its confines.

A widespread practice in the tropics is the use of fodder banks, trees and shrubs closely planted about one to two feet apart that are coppiced one or more times every year, the leaves fed to livestock. At both Las Cañadas and ECHO, I've seen examples of coppiced woody plants such as moringa, katuk, and chaya producing leaves for human consumption. These systems can be very productive because the mature root system of the tree provides lots of energy for quick resprouting and rapid growth of tender young leaves.

Though our woody leaf-crop species are not quite as glamorous as what we've seen in Mexico and Florida, Jonathan and I decided to try this model of food production. One of the benefits of this practice in colder climates is that the leaves on coppiced leaf crops remain tender and edible far into the summer, long after most perennial vegetables have wrapped up their season, at the end of May. In an area that was an overgrown mishmash of water celery, comfrey, sweet cicely, and assorted other odds and ends, we began planting our human fodder bank. We started with littleleaf linden, which Martin Crawford recommends as an edible leaf crop in *Creating a Forest Garden*. I've consumed lots of linden leaves over the years—unlike many perennial vegetables, they can be eaten raw—and find them perfectly palatable, if not remarkably delicious.

To our established lindens, we will add a variety of edible-leaf mulberry. Most mulberry leaves taste pretty terrible, but they are as high as 30 percent protein, and I learned from an unforgettable Mexican lasagna at Las Cañadas that there are some fine varieties for eating, though they must be cooked. We'll also be trying the fragrant spring tree, which is cultivated by Amherst Chinese, a local restaurant that serves the leaves in frittatas. Its strong garlic flavor is an acquired taste, but we are going to give it a try with a flaming magenta variety called Flamingo.

We may also try a few edible-leaf gojis, which I like, though no one else in the house cares for their leaves. To ensure high production of all these leaf crops, we want to have a nitrogen-fixer in the mix, preferably one that coppices just like the other members of the polyculture. Many nitrogen-fixers are poisonous, however, and we wouldn't want any mixups like our daffodil–garlic chive confusion. Bush clover seems to be the right species for the job. Though the leaves are no more delicious than red or white clover, they are non-toxic and considered edible. I have found them to be a serviceable if not gourmet food. At least they are not going to poison us if they end up in the stew by accident. And bush clover grows vigorously in our urban soils and can be coppiced multiple times a year for mulch.

Underneath our coppiced trees and shrubs we will grow some kind of shade-tolerant groundcover, with luck something that provides some pest control, edibility, or other useful function. It may seem strange that we are planting more edible leaves, given that we already have too many. But I like linden and mulberry leaves, and we would be glad to have more perennial leaf crops in summer. Whatever surplus we have would make great rabbit or tilapia food, as we are hoping to add more microlivestock. But even if we had all the leaves we could ever want, neither Jonathan nor I could ever resist the allure of this powerful mix of new crops and a new perennial production model.

29

CHECKING BACK IN AFTER EIGHT SEASONS

Looking back, I find it almost scary to see how closely Jonathan and I were able to achieve our goals. With ideas and support from our families, we were able to find a duplex with a shared garden. Against all likelihood, we both met sweeties who moved in within a month of each other. Jonathan and I have stayed buddies and kept working on the garden, even though we each have our own side of the duplex now. Both of us were able to leave our day jobs in part because of what we're learning here in the garden. It's enough to make me feel cautious about setting my next round of goals: better be careful what I wish for!

From the beginning, Jonathan and I viewed our time in Holyoke as an experiment to test the hypothesis that Dave and I put out while writing *Edible Forest Gardens*. While a lot of individual experiments have failed, overall there is no doubt in my mind that cold-climate forest gardening is a model that can work.

A multistoried forest garden in Massachusetts can produce food from trees, shrubs, herbs, and fungi even in the shade. There are many fine low-maintenance perennial food plants, and with some refinement and observation they can be grown in successful polycultures. Useful native plants can play a valuable role in edible landscapes. Ponds can grow food, and asphalt can be a boon to growing tropical crops in the North. Microlivestock can be managed in a healthy way to improve the garden without intruding on neighbors. More broadly, it is possible to garden a whole yard and to turn terrible soil into an edible oasis.

In 2011 we finally got around to testing our lead levels again and found that our strategies of mulching, liming, and increasing organic matter had indeed moved our lead levels from "low" to "no danger." Our mineral nutrients were still low, but our organic matter had gone up by almost 500 percent.

Harder to nail down are the emergent properties of a garden that has come to life. We do this by looking at the balance of pest populations, the active presence of fungi and other life in our soil, and the spontaneous arrival of local beneficial wildlife, from birds to salamanders. Initially our yard scored fair using the *Edible Forest Gardens* habitat assessment scheme. Our lack of open water, nectary plants, bird food, and lumpy texture (habitat diversity) resulted in little opportunity for beneficial insects and birds. Today our garden features a pond, full-season nectary plants, and improved bird food availability. We have outdone ourselves on lumpy texture. The number and diversity of bird species visiting the garden continues to increase; this year we had an ant-eating flicker in residence for the first time. Our ecosystem assessment score has increased to between good and excellent. We can't yet get up to the highest ranking because to do so we would need to be part of a larger, contiguous healthy ecosystem. Should more of our neighbors adopt our gardening style, perhaps we could achieve that some day.

Another way to evaluate our success is the number of garden tours we give in a year and the way people's eyes light up when they see pictures of our garden when we are out on the road teaching. Part of me feels embarrassed by this attention; I tend to focus on all the mistakes we have made and work we have yet to do. But having seen so many of the experiments people are working on, I've come to appreciate that while we've made our mistakes, we have done a lot of things right. Although I hope this book will inspire others to garden their whole yards and create little pieces of paradise, it can be done so much more simply than we have chosen to do. Our desire to try many new things—new models of production, hundreds of new and interesting species—meant that we put a lot more time into a garden of this size than any reasonable person would ever do. It can be as simple as planting out some American persimmons or pawpaws or putting up a little greenhouse. My hope is that in a decade or two the work Jonathan and I have done in our garden will be an anachronism, a stumbling attempt that overlooked many obvious things that could be done on a piece of land like this. We are easily a step up from chemical lawns or small annual vegetable gardens, but compared to the potential of this kind of gardening and where it seems to be going, I think we'll look like dinosaurs.

PERMACULTURE GREENHOUSE REALIZED

by Jonathan Bates

Climate change is giving us a lot of opportunities to respond creatively to disturbance. I think the real short-term danger of climate change is not slightly warmer temperatures but the greater intensity and frequency of extreme weather. In 2011 we saw a hurricane, tornadoes, and a freakishly heavy October snowstorm pass through our county in Massachusetts. Our little backyard escaped the first two but was devastated by Snowtober 2011. The trees had not yet shed their leaves, and in fact many had yet to change color when the snowstorm came. Heavy, wet snow clung to the leaves and brought branches down throughout our neighborhood. We heard branches breaking every couple of minutes all night.

Meg, Marikler, Eric, and I were standing at a second-story window looking out at the backyard that night when an enormous branch from our neighbor's Norway maple came crashing down and flattened our greenhouse. We groaned, realizing there would be no fresh salads or warm, sunny sanity provided by our greenhouse that winter. Eric and I spent about two days feeling demoralized before we got excited about the possibility of building a new greenhouse.

There was the scream of a saw ripping through sixteen feet of two-by-eight spruce lumber. At five-eighths-inch thick, these rips were to be the backbone of twelve curved trusses. We molded them with a borrowed antique jig and spliced the trusses together tightly with short blocks of two-by-fours between the curved lengths of spruce lumber rips. Along with two friends, I fabricated twelve of these arched trusses in a day. They became the main support components of the greenhouse, which today stands eleven and a half feet in the air.

Why did we build such a unique and seemingly complex structure? Why not purchase a high tunnel greenhouse or a kit like we did the first time? I was inspired by Steve Breyer's insulated greenhouses at Tripple Brook Farm. Steve's design works—and requires significantly less material and money than most greenhouse designs. With no supplemental heat, his greenhouse creates a plant hardiness zone similar to northern Florida in the winter. In addition to winter vegetables, it is warm enough to grow pomegranates, figs, hardy citrus and kumquats, and Chilean guavas. And with a sophisticated, but low-tech aquaponics production model, we would be able to grow fish in tanks, their waste running through several living ecosystems producing plant foods for people and the fish themselves. More animal protein would be very welcome in our backyard cuisine.

An all-wooden skeleton, simple in design and constructed using reclaimed or recycled materials with the help of volunteers, made our new greenhouse inexpensive to build. For the foundation, we used old iron pipe from Meg's family's farm. Part of the frame was made with donated wood from the local home improvement center that salvages building materials. We used recycled billboard vinyl for the east, west, and north outer walls. Eric and I rented an eleven-foot box truck for collecting reclaimed insulation—one hundred and sixty four-by-four sheets of one-inch foam insulation installed in four layers to protect our new greenhouse from the cold of winter.

I shared the construction experience with dozens of people during multiple workshops and building days. On Easter weekend 2012, we invited Meg's father and stepmother and my mom and dad to enjoy a weekend of building with us. Working alongside my dad, who was inspired by such a grand project, and Meg's father, who has confidence and skill as a builder along with a quiet humility, meant that we created something much more than a greenhouse. The months of gathering materials, organizing volunteers, and the sweat and exhaustion was all worth it for the connection of working with others on a project like this, for the skills and knowledge I gained along the way, and for that first bite of a Massachusetts-grown avocado.

Today our permaculture greenhouse (or bioshelter) sits in the back of the garden like a cathedral. Its arches don't look strong, but they can manage the weight of deep winter snows. With a triple, clear-plastic south face, the remainder is wrapped in thick insulation, giving it a subtropical internal climate. It stands as a monument to good design and the potential of abundance.

30

WHAT'S STILL ON THE LIST?

Jonathan and I maintain a list of garden projects broken down into a seasonal calendar. We use a small whiteboard in the toolshed to keep track of short-term projects and a bigger flipchart in my basement office. We also keep notes all season on what we want to do the following year. Besides the usual transplanting and fiddling around, we intend to continue systematically reworking patches of our gardens that aren't up to snuff and installing more next-generation polycultures.

Earlier this year we dismantled our compost bins. Our chickens do such a good job that the few things they won't eat can be handled in a fifty-gallon compost drum. In the compost-rich area where once our bins resided, we hope to plant several new polycultures. This area should have excellent fertility, though it is in partial shade from the neighbor's garage for much of the year. The new polycultures we want to plant here are sunchoke–hog peanut, Chinese artichoke–bush clover, and elderberry–groundnut–earth chestnut.

Jonathan and I also want to experiment more with grafting. Quince, medlar, and hawthorns can all be grafted onto each other. This year while teaching in California, I had the opportunity to try the Aromatnaya Russian quince. While most quinces are aromatic, they are also solid and must be cooked. Aromatnaya, however, can be eaten raw and may be the closest thing to a guava that can be grown outdoors in Massachusetts. Medlars, which have a flavor like cinnamon applesauce, and the many species of sweet-fruited hawthorns, like Arnold's hawthorn, would also be a welcome addition to our late fall fruit mix. Because all three species can fruit in partial shade, we will be planting them where our compost pile once stood and playing around with grafting them onto each other.

In the shady areas on the north side of the house, we will be experimenting with polycultures for full shade, using species like giant ostrich fern, mayapple, and a range of native groundcovers. We also hope to address the two weeks in August that have little fruit by planting blackberries, black raspberries, and maybe hybrid marionberries in the rhizome barrier beds where we trialed our sunchoke polycultures.

We are getting tired of spreading wood chips on our paths every few years and hope to once again follow in Martin Crawford's footsteps by experimenting with low, clumping, shade-tolerant grasses in our pathways. We will trial some species to see which can stand foot traffic and won't spread into our beds. We anticipate needing to cut them back when they start to seed. We could feed the clippings to chickens or, if we add them, rabbits.

Rabbits have been on our list for a long time. It is possible to raise rabbits with no purchased feed if they get enough vegetation from the garden. Our thought is to buy some meat rabbits every spring and raise them on weeds until fall, when they would become savory stew. We'd like to build the rabbit hutches inside the chicken run. Beneath the cages, where their wonderful manure falls, we'd install worm bins. This will give us a yield of worm castings and occasional harvests of worms to feed to our chickens.

Worms aren't the only invertebrate livestock we are thinking of incorporating. Our friend Lisa DiPiano keeps bees and has offered to locate a hive on the roof of our toolshed. She would care for them and teach us about raising bees. By keeping the bees up high, we would avoid their bothering neighbors or stinging anyone who came by on a tour. If this doesn't work out, we'll try native orchard mason bees, which do not make honey but are easy to raise and are fantastic pollinators.

Another insect on our list is the black soldier fly. These are coming into vogue in permaculture circles for their ability to turn kitchen scraps into compost and maggots. The idea of raising maggots on purpose is counterintuitive, but I developed a warm spot in my heart for writhing, wriggling insect larvae after seeing Marikler's success with silkworms. Soldier fly larvae don't smell bad, and their texture is dry and firm rather than soft and slimy like housefly larvae. Some of them showed up in our compost pile on their own, and they didn't bother me at all.

In fact, soldier fly larvae give off hormones that prevent other kinds of flies from colonizing the compost. Commercially available soldier fly larvae production units like the BioPod take advantage of the larvae's instinct to crawl up and out of the pile before metamorphosis by capturing the escaping larvae in a jar. The larvae make a fantastic feed for chickens and fish. As Harvey Ussery points out in *The Small-Scale Poultry Flock,* real sustainability in poultry production means getting away from bagged feed and producing as much chicken feed on-site as possible. Feeding larvae that were fed from the waste stream, as opposed to feeding grains that people could themselves eat, is simply a sensible idea.

In the tropical garden out front, Jonathan and I are going to experiment with some marginally hardy relatives of ginger (USDA zones 6 and 7). These shade-loving edible ornamentals might like conditions under our bananas. We are ordering zedoary, cultivated for its starchy roots as a shade crop in the tropics; mioga ginger, with edible shoots and flowers; and edible-flowered *Alpinia japonica* and *Hedychium coronarium.*

Jonathan and I are working on a project to breed "neohybrid" perennial broccoli by crossing sea kale and a number of its relatives. Our hope is to develop sea kale with a single large head instead of lots of little ones like we have now, using simple techniques we learned from Badgersett Research Corporation in Minnesota. This year we may be able to save seed from the first of our crosses, and we plan to add even more related perennial broccolis to the genetic blend.

I've already mentioned that we hope to graft Chinese chestnut onto our shrubby chinquapins, but another chestnut opportunity has come our way. One of our neighbors' Norway maples was so badly damaged in the October snowstorm that it lost half of its branches.

Although we are not thrilled that some of those branches destroyed our greenhouse, we would be happy to see this tree go. We recently spoke with our neighbors, who said we're welcome to take it down, which would open up light on the north edge of our garden. We'll plant the biggest chestnut we can find there, and we plan to graft several varieties onto it.

Jonathan wrote about our new greenhouse. We are realizing a dream we have held for a long time. We've already hosted some workshops on greenhouse design and construction. Who knows? Maybe we will even get around to installing that irrigation system.

One of the strangest regrets I have about our garden is that we don't have more people helping out. If each side of the duplex had another couple living in it, we could easily grow more produce than eight people could eat. With Jonathan and me as the main gardeners and Meg and Marikler as dessert chef and livestock queen, we are close to maxing out our available time. Jonathan points out that the more of our income comes from the garden, the more time we can give it. Indeed, if we had fewer hours of day jobs and other responsibilities, the four of us could do the work of eight or even twelve backyard dilettantes.

The annual beds, for example, could be so much more productive if someone had time to manage them biointensively. Every one of our annual beds could have a cold frame to extend the season into spring and fall. Jonathan could get more aggressive about mushroom production with inoculated mulch on every bed and stacked mushroom logs in the shade of structures and fruit trees. Marikler could tend tiny herds of microlivestock.

We could be much more systematic about collecting neighborhood leaves and yard waste and even collect compost from our neighbors or local restaurants. If it were legal, we could raise a dwarf pig every year on neighborhood food scraps and have a pig roast in the backyard every fall for the neighbors who'd provided us with their kitchen waste.

If it were legal, we could also use our household greywater by filtering it through a small wetland and then irrigating with it. The wetland could provide lots of mulch material and livestock feed. As long as we are fantasizing about city policy changes, a composting toilet could allow us to return the nutrients we routinely remove from the garden and flush away into the already polluted river. As I was finishing this book, there were some promising developments. The fall 2011 city election saw the victory of Alex Morse, a young candidate with some progressive ideas. A number of Hispanic candidates won seats on the city council, and my friend Lisa Rodriguez-Ross, the wife of my friend Daniel of Nuestras Raíces, was appointed our new city solicitor. Holyoke may be poised on the brink of a new renaissance if it can welcome green development in a way that includes the Hispanic community.

Of course even under the most ideal circumstances, we would still not be producing all of our food—really just a lot of produce and some protein. Today we buy bread and dairy products and other foods like bananas and avocados. But our goal was never self-sufficiency, and I don't think that's really realistic on a plot of this size.

A ring of farms outside a small city like Holyoke, however, could provide the wheat, beans, meats, and dairy products to supplement produce grown intensively close to home. Neighborhood community gardens, edible landscaping and parks, urban farm blocks, and greenhouses could produce food by the ton. Martin Crawford provided hundreds of nut trees to become the new street trees for the town of Totnes in England. Totnes has engaged in a communitywide process known internationally as the "transition town" movement, which works to address climate change and peak oil by increasing local resilience, reducing emissions, and sequestering carbon. Sometimes I drive around and look at the ornamental pears that line many of our streets and think of our friend Justin West, Martin Crawford's protégé (now running a nursery in New Jersey), who has been grafting them over to select varieties of European and Asian pear. A guerrilla grafting campaign could turn many city streets and parking lots into pear-producing paradises in just a few years.

Say the economy continues to degrade or the price of petroleum rises dramatically as it gets more and more expensive to extract or other social and environmental costs of our way of life that have been externalized finally and truly hit home. It may be that having lots of people share relatively small units of urban housing and produce astounding amounts of food on them makes a lot more economic sense. I find it comforting to know that it can be done.

One of the principles in *Permaculture: A Designers' Manual* that has always inspired me is that "the yield of a system is theoretically unlimited. The only limit on the number of uses of a resource possible within a system is in the limit of the information and the imagination of the designer." Our experience reinforces this idea and refutes the notion of scarcity.

Actually, our fruit and nut trees have yet to even mature. Our persimmon could grow another twenty feet. Our kiwis haven't even fruited, and our hazelnuts are just starting to bear well. The edible-leaf tree crops we're planting next year should dramatically increase our midsummer perennial vegetable production. Even if we never have more farming housemates, we seem to be on a path of ever-increasing yields.

Backyard produce is not sufficient to feed a city, but it can make a substantial impact. Most of what we consume on a daily basis is carbohydrates, in the form of bread, pasta, rice, and tubers like potatoes. What if we planted our streets with staple crop-producing trees like chestnuts? Our block is about two hundred feet long. By my calculation, there is room for a single, thirty-five-foot-wide row of chestnuts on each side of the street, which is about a third of an acre. Chestnuts might reasonably yield twenty-four hundred pounds per acre in less than ideal urban circumstances, or eight hundred pounds on our block. Of this, six hundred and eighty pounds are edible, with 15 percent of the weight being shells. The trees would shade parked cars and front yards, even if cleaning up all of those husks would be a challenge. (Perhaps they could be made into charcoal for use in roasting the neighborhood pig.)

How does that compare to loaves of bread? The whole-wheat loaf we have in the fridge today has sixteen slices at twenty grams of carbohydrate each for a total of three hundred and twenty grams per loaf, or 0.7 pounds of carbohydrates per loaf. Chestnuts are 78 percent carbohydrates, so our street's production of carbohydrates would be five hundred and thirty pounds. That's the equivalent of over seven hundred and fifty loaves, or thirty-seven loaves of bread for each of the twenty family units on our block. Not bad for a two-hundred-foot stretch of residential street!

The list of potential projects for our yard and our neighborhood is endless. Every year Jonathan and I find new things to get excited about, and there is no end of new species and techniques to discover. Even only halfway knowing what we were doing, we have increased the beauty, ecosystem health, and food production of what was once an empty lot. We made our little paradise here. Imagine what would happen if we as a species paid similar attention to all the degraded and abandoned lands of the world.

EPILOGUE

When Jonathan and I moved to the house in Holyoke, we thought it was possible we might spend the rest of our lives here. There's so much in our garden that could keep us interested for a long time. And the longer we stay and the bigger our fruit trees get, the harder it is to think about leaving. But the four of us in the house agree that our days here are probably numbered.

Jonathan and Megan's nursery is growing, and at some point they will need a lot more space. Megan's father has a farm near Ithaca, New York, with family nearby. Marikler and I have a plan as well. In the fall of 2011, our son, Daniel Benjamin, was born. I joked with Jonathan that Daniel was my main propagation project for the year. Marikler's long-term goal is to work in international sustainable development again. That means a master's degree alongside raising Daniel and eventually a move to some place that offers her career potential. As long as I'm near an airport, I can write and teach from anywhere.

We all feel the allure of a larger piece of land. I feel as if I have learned a lot of what I want to learn about small-scale backyard gardening, and I've proven to myself that the edible forest gardens model works at this scale. But I can't get the idea of the ten- or fifteen-acre farm out of my head. I imagine large-scale rainwater harvesting strategies, alternating strips of chestnuts, persimmons, and organic no-till crops. I picture being able to use sheep and hogs and poultry to manage the understory of a large-scale carbon-sequestering food forest. That's where my fantasies are running these days, although it may be that my hopes of helping to steer climate change funds into perennial agriculture projects around the world means that I spend more time at a desk than in my backyard. Lord knows my back is in no shape for a commercial-scale farming enterprise.

So what will happen to our garden? We hope we'll find someone who will be excited to move into a house with kiwis and persimmons in the yard. I think there's enough challenge here that someone wouldn't feel constrained by the choices Jonathan and I have already made. But should worse come to worst and it ends up getting sold to someone who just wants to pave it all over and put up a giant swimming pool in the back, this garden has replicated itself many times over already. The offspring of our plants are growing in hundreds of gardens. Through the Apios Institute wiki, other people are replicating the best of our polycultures, and many have used the *Edible Forest Gardens* design process that we first tested here.

Writing the story of your own garden is a bit of an act of hubris and not something I would have come up with on my own had someone not suggested to me that it was a story worth telling. Writing this book has forced me to look beyond the things we did wrong and those still on our list and appreciate the successes of what we've done here. Now I stop a little longer during my daily walks around the garden and think back on how far we've come. In fact, I think I'll go outside and have a persimmon right now.

ACKNOWLEDGMENTS

First and foremost, I'd like to acknowledge Jonathan. Besides being a full partner in our garden design, he did the lion's share of its implementation. Without his steady work, I'd probably still be writing about plants from a gardenless apartment somewhere. His design and management ideas, construction abilities, and intensive research on plants, fungi, and microlivestock have taught me much. He's also a reliable and trustworthy friend, not always easy to find in this world. It's hard to believe we've gardened together for twelve years.

I'm very grateful to Gwen and Ralph Bates for allowing us to rent our home before we bought it. This generous act made our experiment possible.

Thanks to my agent, Kit Ward, for suggesting we write the story of this garden. It would not have occurred to me, but I've really enjoyed much of the process and think it has real value. Brianne Goodspeed at Chelsea Green believed in this book and guided us through a major rewrite. She always understood our vision and worked diligently to bring it out in a way that readers could more easily take in. Margo Baldwin, Chelsea Green president and publisher, is a great supporter of permaculture in general, and I'm grateful to her for getting behind my projects in particular.

Everyone at Chelsea Green was very patient with my lapses in availability. We knew I'd take a paternity leave after the birth of my son, Daniel, but no one expected I would turn out to have Crohn's disease and need major surgery in the middle of rewriting.

I hope readers can see that Marikler is a delight to live with. I've counted on her throughout the process of writing, parenting, and surgery. My thanks to her for her love and patience. Daniel, someday you'll be able to read this and know what I was busy with down in the basement the first ten months of your life.

Eric Toensmeier

I would like to start by praising Eric for his decade of mentorship and garden companionship and for seeing in me more then I saw in myself sometimes. Without his encouragement and support, the garden, and thus this book, would never have been created. The agrogeekosystem we have will always be a source of vibrant inspiration to me.

A lover of life, Meg has inspired me to live more fully and love completely. Once upon a time I directed my passion into the garden in the hope that the flowers would one day bear fruit. Now my passion for life is interdependent with and expressed through my love for Meg. Our experiences together bear the fruit of life: connection, understanding, kindness, sharing, laughter, closeness—each of which expands through time. I look forward to continuing the paradise we have created.

It is difficult to put into a few words the life my parents have fostered around me. This book is just one leap into the great unknown they have supported. For listening to my stories, offering advice, cheerleading my choices, and helping where they could financially, I am deeply and infinitely grateful to them. They are the most exceptional parents a son could have. Thanks for bringing me into this world, Mom and Dad.

And finally, I dedicate my work in the garden and this book to the future generations: Daniel Toensmeier and my son, Jesse Bates. They will be the ones who help create a world where all beings may thrive.

Jonathan Bates

APPENDIX A:
DESIGN PLAN AND FIELD SKETCHES

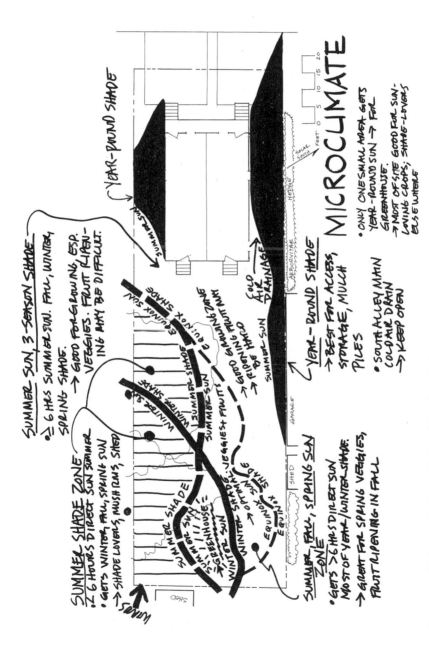

Patterns of light and shade. Field sketch by Dave Jacke.

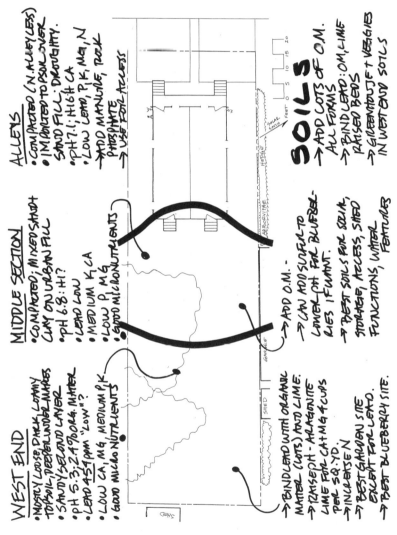

WEST END
- MOSTLY LOOSE DARK LOAMY TOPSOIL, DEEPER UNDER MAPLES
- SANDY SECOND LAYER
- PH 5.3; 2.4% ORG. MATTER
- LEAD 459 ppm "LOW"?
- LOW CA, Mg, MEDIUM P, K
- GOOD MICRO NUTRIENTS

→ BIND LEAD WITH ORGANIC MATTER (LOTS) AND LIME.
→ 12 in SEPT - AGLADONITE LIME FOR CA + Mg 4 CUPS PER SQ. YD.
→ INCREASE N
→ BEST GARDEN SITE EXCEPT FOR LEAD.
→ BEST BLUEBERRY SITE.

MIDDLE SECTION
- COMPACTED; MIXED SAND+ CLAY ON URBAN FILL
- PH 6.8; HI?
- LEAD LOW
- MEDIUM K, CA
- LOW P, Mg
- GOOD MICRONUTRIENTS

→ ADD O.M. -
→ CAN ADD SUN FILL TO LOWER O.M. FOR BLUEBERRIES IF WANT.
→ BEST SOILS FOR SOLAR, STORAGE, ACCESS, SHED FUNCTIONS, WATER FEATURES

ALLEYS
- COMPACTED (N. ALLEY LESS)
- IMPORTED TO PSOL OVER SAND FILL - DROUGHTY.
- PH 7.1; HIGH CA
- LOW LEAD, P, K, Mg, N

→ ADD MANURE, ROCK PHOSPHATE
→ USE FOR ACCESS

SOILS
→ ADD LOTS OF O.M. ALL FORMS
→ BIND LEAD: O.M., LIME RAISED BEDS
→ GREENHOUSE + VEGGIES IN WEST END SOILS

Our three types of terrible soil: sterile sand-gravel fill, compacted clay with "urbanite" rubble, and acid sand with lead. Field sketch by Dave Jacke.

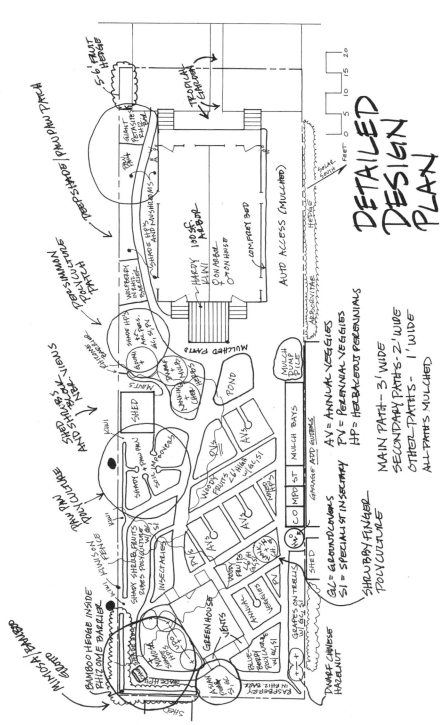

Our master permaculture design for the garden. Sketch by Dave Jacke.

APPENDIX B: PLANT SPECIES BY LAYER

These are species we have successfully grown and harvested. All have overwintered except a few of the aquatic and tender tropical species. Note that "native" means native to the Northeast and not necessarily to Massachusetts or our county.

Trees and Bamboos over 12'		
Common Name	Botanical Name	Features
mimosa Rosea	*Albizia julibrissin*	nitrogen-fixing, hummingbird nectar
pawpaw Rappahannock, Shenandoah, and unnamed seedlings	*Asimina triloba*	edible fruit, native
American persimmon Early Golden and unnamed male	*Diospyros viginiana*	edible fruit, native
yellowgroove bamboo	*Phyllostachys aureosulcata*	edible shoots, screen, poles
smoothsheath bamboo	*Phyllostachys nidularia*	edible shoots, screen, poles
nudesheath bamboo	*Phyllostachys nuda*	edible shoots, screen, poles

Trees, Shrubs, and Herbs 6–12'		
Common Name	Botanical Name	Features
beach plum Nana	*Prunus maritima*	edible fruit, native
dwarf hazelnut	*Corylus* spp.	nuts
false indigo	*Amorpha fruticosa*	nitrogen-fixer, native
goumi Sweet Scarlet	*Eleagnus multiflora*	edible fruit, nitrogen-fixer
Japanese fiber banana	*Musa basjoo*	leaves for tamales and food wraps
mulberry Geraldi dwarf	*Morus macroura*	edible fruit, silkworm food
semidwarf Asian pear Chojuro and other varieties	*Pyrus* spp.	edible fruit
sunchoke, Jerusalem artichoke	*Helianthus tuberosa*	edible tubers, native, nectary plant
udo	*Aralia cordata*	edible shoots, fantastic nectary plant

Shrubs and Herbs 3–6'

Common Name	Botanical Name	Features
asparagus Purple Knight	*Asparagus officinalis*	edible shoots
black currant Consort	*Ribes nigrum*	edible fruit
bush clover	*Lespedeza bicolor*	nitrogen-fixer
clove currant Crandall	*Ribes odoratum*	edible fruit
cow parsnip	*Heracleum maximum*	native, nectary, minor edible stalks and roots
giant fuki	*Petasites japonicus giganteus*	edible stalks, nectary
giant Solomon's seal	*Polygonatum commutatum*	edible shoots, roots, native
goji leaf	*Lycium chinense*	edible leaves, fruit
halfhigh blueberry	*Vaccinium* spp.	edible fruit, native
honeyberry, haskap	*Lonicera kamschatica*	edible fruit
jostaberry	*Ribes* spp.	edible fruit
juneberry Regent	*Amelanchier* spp.	edible fruit, native
Kirilow's indigo	*Indigofera kirilowii*	nitrogen-fixer
littleleaf linden (coppiced)	*Tilia cordata*	edible leaves
ostrich fern, giant ostrich fern	*Matteuccia struthiopteris*	edible shoots, native
red alder (frost coppiced)	*Alnus rubra*	nitrogen-fixer
red and yellow raspberry	*Rubus idaeus*	edible fruit
red currant Red Lake, Blanca, Pink Champagne	*Ribes rubrum*	edible fruit
Russian comfrey Bocking 4, Bocking 14	*Symphytum × uplandicum*	mulch plant, beneficial insect habitat
stinging nettle ("stingless")	*Urtica dioica*	edible leaves (cooked!)
tiger lily	*Lilium lancifolium*	edible bulbs
Turkish rocket	*Bunias orientalis*	edible broccolis, leaves

Woody Vines

Common Name	Botanical Name	Features
akebia	*Akebia quinata*	edible fruit
arctic kiwifruit	*Actinidia kolomitka*	edible fruit
grape (*Labrusca* hybrids) Glenora, Reliance	*Vitis hybrids*	edible fruit
hardy kiwifruit	*Actinidia arguta*	edible fruit

Herbaceous Vines

Common Name	Botanical Name	Features
Chinese yam, cinnamon vine	*Dioscorea opposite*	edible roots, aerial "yamberries"
cucumber berry	*Melothria pendula*	edible minicucumbers, native
groundnut including Nutty	*Apios americana*	edible tubers, native, nitrogen-fixer
maypop	*Passiflora incarnata*	edible fruit, native
perennial wild bean	*Phaseolus polystachios*	edible beans, native, nitrogen-fixer
red hailstone vine	*Thladiantha dubia*	edible shoots
sprawling spinach	*Hablitzia tamnoides*	edible shoots, leaves

Shrubs and Herbs 18–36"

Common Name	Botanical Name	Features
anise hyssop	*Agastache foeniculum*	tea
Chinese fairy bells	*Disporum cantonense*	edible shoots
clammy ground-cherry	*Physalis heterophylla*	edible fruit, native
common milkweed	*Asclepias syriaca*	edible shoots, leaves, flowers, pods, native
daylily	*Hemerocallus fulva*	edible shoots, flowers, roots, groundcover
French tarragon	*Artemisia dracunculus*	culinary
garlic chives	*Allium tuberosum*	edible leaves, flowerbuds, groundcover
gooseberry Hinnomaki Red, Invicta, and more	*Ribes uva-crispa*	edible fruit
hayscented fern	*Dennstaedtia punctilobula*	native, groundcover
heartleaf aster	*Aster cordifolia*	nectary, native
Illinois bundleflower	*Desmanthus illinoiensis*	edible beans, nitrogen-fixer, native
longleaf ground-cherry	*Physalis longifolia*	edible fruit, native
multiplier onion	*Allium cepa aggregatum*	edible bulbs
New England aster Purple Dome	*Aster novae-angliae*	nectary, native
New Jersey tea	*Ceanothus americanus*	tea, nitrogen-fixer, native
perennial leek Kurrat, Elephant Garlic	*Allium ampeloprasum*	edible leaves and bulbs

running juneberry	*Amelanchier stolonifera*	edible fruit, native
sage	*Salvia officinalis*	culinary
scorzonera	*Scorzonera hispana*	edible leaves, shoots, roots
sea kale	*Crambe maritima*	edible broccolis, shoots, leaves
sessile-leaved tick trefoil	*Desmodium sessilifolium*	nitrogen-fixer, native
skirret	*Sium sisarum*	edible leaves, roots, nectary
sweet cicely	*Myrrhis odorata*	edible seeds, leaves, roots, nectary
sweet goldenrod	*Solidago odora*	tea, nectary, native
sylvetta perennial arugula	*Diplotaxis muralis*	edible leaves
walking onion	*Allium cepa proliferum*	edible scallions, topset onions
Welsh onion	*Allium fistulosum*	edible scallions
wild garlic	*Allium canadense*	edible leaves and bulbs, native
wood nettle	*Laportaea canadensis*	edible shoots (cooked!), native
yarrow	*Achillea millefolium*	nectary, groundcover, beneficial insect habitat

Shrubs and Herbs 6–18"

Common Name	Botanical Name	Features
chives Profusion	*Allium schoenoprasum*	culinary, native
ramps	*Allium tricoccum*	edible leaves, bulbs, native
wild chervil Ravenswing	*Anthriscus sylvestris*	culinary, nectary
silverweed	*Argentina anserina*	groundcover, edible roots
licorice milk vetch	*Astragalus glycyphyllos*	nitrogen-fixing, groundcover
earth chestnut	*Bunium bulbocastanum*	edible leaves, tubers, nectary
camas	*Camassia quamash*	edible bulbs
wild hyacinth	*Camassia scillioides*	edible bulbs, native
good King Henry	*Chenopodium bonus-henricus*	edible shoots, leaves
perennial chickpea	*Cicer microphyllum*	edible beans
edible hosta	*Hosta sieboldiana*	edible shoots (who knew!)
"Chameleon"	*Houttuynia cordata*	culinary, groundcover
perennial wild lettuce	*Lactuca canadensis*	edible leaves, flowerstalk, native
osha	*Ligusticum porteri*	nectary

Scotch lovage	*Ligusticum scoticum*	edible leaves, nectary
fernleaf biscuitroot	*Lomatium dissectum*	edible roots, nectary
birdsfoot trefoil	*Lotus corniculatus*	nitrogen- fixing, groundcover
mallow Gumbo Leaf	*Malva moschata*	edible leaves
lemon balm	*Melissa officinalis*	tea
peppermint	*Mentha piperita*	tea, groundcover
spearmint	*Mentha spicata*	tea, groundcover
salad mint	*Mentha* spp.	tea, groundcover
water celery	*Oenanthe javanica*	edible leaves, groundcover, nectary
native pachysandra	*Pachysandra procumbens*	evergreen groundcover, native
mayapple	*Podophyllum peltatum*	edible fruit, native
prostrate sand cherry Paw-nee Buttes, Select Spreader	*Prunus besseyi*	edible fruit, native, groundcover
hyssop-leaved mountain mint	*Pycnanthemum hyssopifolium*	tea, nectary, native
short-toothed mountain mint	*Pycnanthemum muticum*	tea, nectary, native
sorrel Profusion	*Rumex acetosa*	edible leaves, groundcover, mulch plant
bloody dock	*Rumex sanguineus*	edible leaves
silvershield sorrel	*Rumex scutatus*	edible leaves, groundcover
Chinese artichoke	*Stachys affinis*	edible tubers, groundcover
dwarf comfrey	*Symphytum grandiflorum*	mulch plant, groundcover, beneficial insect habitat
dwarf comfrey Hidcote Blue	*Symphytum* spp.	mulch plant, groundcover, beneficial insect habitat
red clover	*Trifolium pratense*	nitrogen-fixer
lowbush blueberry	*Vaccinium angustifolium*	edible fruit, native

Shrubs and Herbs up to 6"

Common Name	Botanical Name	Features
alpine strawberry	*Fragaria vesca alpina*	edible fruit
barren strawberry	*Waldsteinia fragaroides*	evergreen groundcover, native
common blue violet	*Viola sororia*	edible leaves and flowers, groundcover, native

cutleaf toothwort	*Dentaria diphylla*	wasabi-like roots and leaves, native
dwarf coreopsis	*Coreopsis auriculata nana*	nectary, evergreen groundcover, native
dwarf horsetail	*Equisetum sylvaticum*	evergreen groundcover, native
green and gold	*Chrysogonum virginianum*	nectary, evergreen groundcover, native
hybrid groundcover raspberries	*Rubus × stellarcticus*	edible fruit, groundcover
kinnikinnick, bearberry	*Arctostaphyllos uva-ursi*	evergreen groundcover, native
lingonberry	*Vaccinnium vitus-idaea*	edible fruit, evergreen groundcover, native
musk strawberry	*Fragaria moschata*	edible fruit, evergreen groundcover
partridgeberry	*Mitchella repens*	edible fruit (boring), evergreen groundcover, native
prostrate birsdfoot trefoil	*Lotus corniculatus plenus*	nitrogen-fixer, foot-tolerant groundcover, sterile
saffron crocus	*Crocus sativus*	the cultivated saffron
silver shield sorrel	*Rumex scutatus*	edible leaves, groundcover
strawberry	*Fragaria anassana*	edible fruit, evergreen groundcover
sweet violet	*Viola odorata*	edible leaves and flowers, groundcover
thyme	*Thymus vulgaris*	culinary
violet Rebecca	*Viola* spp.	edible leaves and flowers (excellent)
white clover	*Trifolium repens*	nitrogen-fixer, foot-tolerant groundcover
wild ginger	*Asarum canadense*	groundcover, native
wild strawberry Intensity	*Fragaria vesca*	edible fruit, evergreen groundcover, native
wild strawberry Kelly's Blanket	*Fragaria virginiana*	edible fruit, evergreen groundcover, native
willowleaf sunflower Low Down	*Helianthus salicifolius*	nectary, groundcover, native
wintergreen Very Berry	*Gaultheria procumbens*	edible fruit, evergreen groundcover, native

Self-seeding Annuals and Biennials

Common Name	Botanical Name	Features
annual ground-cherry	*Physalis pruinosa*	edible fruit, native
black nightshade	*Solanum ptycanthum*	edible fruit, cooked leaves, native
borage	*Borago officinalis*	edible flowers and leaves, beneficial insect habitat
corn salad, mache	*Valerianella locusta*	edible leaves
hog peanut	*Amphicarpa bracteata*	edible beans, nitrogen-fixer, groundcover, native
johnny jump-up	*Viola tricolor*	edible leaves
kale Western Front	*Brassica oleracea*	edible leaves
lablab bean	*Lablab purpureus*	edible beans, flowers, nitrogen fixing
lamb's quarters	*Chenopodium album*	edible leaves
miner's lettuce	*Montia perfoliata*	edible leaves
New Zealand spinach	*Tetragonia tetragonioides*	edible leaves
giant orach	*Atriplex* spp.	edible leaves
purslane	*Portulaca oleracea*	edible leaves, native
Queen Anne's lace	*Daucus carota*	nectary
red shiso	*Perilla frutescens*	culinary
sensitive partridge pea	*Chemaecrista nictitans*	nitrogen-fixer, native
tomatillo	*Physalis ixocarpa*	edible fruit
tomato Yellow Pear	*Lycopersicon esculentum*	edible fruit

Aquatic Species

Common Name	Botanical Name	Features
arrowhead	*Saggitaria latifolia*	edible roots, shoots, native
cattail	*Typha latifolia*	edible shoots, root sprouts, pollen, native
Chinese lotus	*Nelumbo nucifera*	edible roots, seeds
native lotus	*Nelumbo lutea*	edible roots, seeds, native
pickerelweed	*Pontederia cordata*	edible seeds, native
water celery	*Oenanthe javanica*	edible leaves
water mimosa	*Neptunia oleracea*	edible leaves, nitrogen-fixing
water spinach	*Ipomoea aquatica*	edible leaves
watercress	*Nasturtium officinale*	edible leaves

Ornamental Tropical Annuals and Tender Tropicals

Common Name	Botanical Name	Features
edible-leaf hibiscus	*Abelmoschus manihot*	edible leaves
achira edible canna	*Canna edulis*	edible tubers and shoots
ornamental cannas	*Canna* spp.	edible tubers and shoots
luau leaf taro	*Colocasia esculenta*	cooked leaves
chipilín	*Crotolaria longirostrata*	cooked leaves, nitrogen fixing
cranberry hibiscus	*Hibiscus acetosella*	edible leaves
sweet potato	*Ipomoea batatas*	edible tubers, cooked leaves
lablab bean	*Lablab purpureus*	edible beans, flowers, nitrogen fixing
moringa	*Moringa oleifera*	edible leaves, flowers, pods, roots

Winter Salad Greenhouse Favorites

Common Name	Botanical Name	Features
arugula	*Eruca sativa*	edible leaves
Asian brassicas (tat soi, vitamin greens, mizuna)	*Brassica juncea*	edible leaves
beet	*Beta vulgaris*	edible roots, leaves
carrot	*Daucus carota*	edible roots
corn salad, mache	*Valerianella locusta*	edible leaves
miner's lettuce	*Montia perfoliata*	edible leaves
parsley	*Petroselenium crispum*	edible leaves
sorrel Profusion	*Rumex acetosella*	edible leaves
spinach	*Spinacea oleracea*	edible leaves
sylvetta perennial arugula	*Diplotaxis muralis*	edible leaves
water celery	*Oenanthe javanica*	edible leaves
watercress	*Nasturtium officinale*	edible leaves
Welsh onion	*Allium fistulosum*	edible scallions

Ornamental Annuals

Common Name	Botanical Name	Features
bottle gourd	*Lagenaria siceraria*	edible shoots, fruit
chard Bright Lights, beet Bulls Blood	*Beta vulgaris*	edible leaves, roots

eggplant Ping Tung Long	*Solanum melongena*	edible fruit
kale Redbor	*Brassica oleracea*	edible leaves
quailgrass, cockscomb celosia	*Celosia argentea*	edible leaves

RECOMMENDED RESOURCES

BOOKS

Anderson, M. Kat. *Tending the Wild: Native American Knowledge and the Management of California's Natural Resources.* Berkeley: University of California Press, 2005.

Bane, Peter. *The Permaculture Handbook: Garden Farming for Town and Country.* New Society Publisehrs, Gabriola Island, 2012.

Christopher, Thomas, ed. *The New American Landscape: Leading Voices on the Future of Sustainable Gardening.* Portland, OR: Timber Press, 2011.

Coleman, Eliot. *The Winter Harvest Handbook: Year-Round Vegetable Production Using Deep Organic Techniques and Unheated Greenhouses.* White River Junction, VT: Chelsea Green, 2009.

Copyne, Kelly, and Erik Knutzen. *The Urban Homestead: Your Guide to Self-Sufficient Living in the Heart of the City.* London: Process Press, 2010.

Crawford, Martin. *Creating a Forest Garden: Working with Nature to Grow Edible Crops.* Totnes, UK: Green Books, 2010.

Creasy, Rosalind. *The Complete Book of Edible Landscaping: Home Landscaping with Food-Bearing Plants and Resource-Saving Techniques.* San Francisco: Sierra Club Books, 1982.

Cullina, William. *Native Trees, Shrubs, and Vines: A Guide to Using, Growing, and Propagating North American Woody Plants.* Boston: Houghton Mifflin, 2002.

———. *The New England Wildflower Society Guide to Growing and Propagating Wildflowers of the United States and Canada.* Boston: Houghton Mifflin, 2000.

Del Tredici, Peter. *Wild Urban Plants of the Northeast: A Field Guide.* Ithaca, NY: Comstock Publishing, 2010.

Deppe, Carol. *The Resilient Gardener: Food Production and Self-Reliance in Uncertain Times.* White River Junction, VT: Chelsea Green, 2011.

Francko, David. *Palms Won't Grow Here and Other Myths: Warm-Climate Plants for Cooler Areas.* Portland, OR: Timber Press, 2003.

Hemenway, Toby. *Gaia's Garden: A Guide to Home-Scale Permaculture.* White River Junction, VT: Chelsea Green, 2009.

Holmgren, David. *Permaculture: Principles and Pathyways Beyond Sustainability.* Hepburn, Australia: Holmgren Design Services, 2002.

Jacke, Dave. *Edible Forest Gardens.* With Eric Toensmeier. White River Junction, VT: Chelsea Green, 2005.

Jeavons, John. *How to Grow More Vegetables (and Fruit, Nuts, Berries, Grains and Other Crops) Than You Every Thought Possible on Less Land Than You Can Imagine.* Berkeley: Ten Speed Press, 2012.

Kellogg, Scott, and Stacy Pettigrew. *Toolbox for Sustainable City Living.* Cambridge, MA: South End Press, 2008.

Kourik, Robert. *Designing and Maintaining Your Edible Landscape Naturally.* London: Permanent Publications, 2004.

Lancaster, Brad. *Rainwater Harvesting for Drylands and Beyond.* Tucson, AZ: Rainsource Press, 2006.

Martin, Franklin W., Ruth M. Ruberté, and Laura S. Meitzner. *Edible Leaves of the Tropics.* 3rd ed. Fort Myers, FL: ECHO, 1998.

Mollison, Bill, and Reny Mia Slay. *Introduction to Permaculture.* Berkeley: Ten Speed Press, 1997.

Reich, Lee. *Landscaping with Fruit.* North Adams, MA: Storey Publishing, 2009.

Romanowski, Nick. *Edible Water Gardens: Growing Waterplants for Food and Profit.* Flemington, Australia: Hyland House, 2007.

Roth, Susan, and Dennis Schrader. *Hot Plants for Cool Climates: Gardening with Tropical Plants in Temperate Zones.* Boston: Houghton Mifflin, 2000.

Ruppenthal, R. J. *Fresh Food from Small Spaces: The Square-Inch Gardener's Guide to Year-Round Growing, Fermenting, and Sprouting.* White River Junction, VT: Chelsea Green, 2008.

Stamets, Paul. *Mycelium Running: How Mushrooms Can Help Save the World.* Berkeley: Ten Speed Press, 2005.

Thayer, Samuel. *The Forager's Harvest: A Guide to Identifying, Harvesting, and Preparing Edible Wild Plants.* Ogema, WI: Forager's Harvest, 2006.

———. *Nature's Garden: A Guide to Identifying, Harvesting, and Preparing Edible Wild Plants.* Ogema WI: Forager's Harvest, 2010.

Toensmeier, Eric. *Perennial Vegetables.* White River Junction, VT: Chelsea Green, 2007.

Ussery, Harvey. *The Small-Scale Poultry Flock.* White River Junction, VT: Chelsea Green, 2011.

WEBSITES

Agroforestry Research Trust
www.agroforestry.co.uk
Martin Crawford's forest garden information clearinghouse.

Apios Institute for Regenerative Perennial Agriculture
www.apiosinstitute.org
Edible forest gardens and perennial polyculture wiki.

California Rare Fruit Growers
www.crfg.org
Fruit enthusiasts' organization.

Central Rocky Mountain Permaculture Institute
www.crmpi.org
Long-established permaculture training site.

Desert Harvesters
www.desertharvesters.org
Sonoran desert edible landscape organization.

Las Cañadas Centro de Permacultura
www.bosquedeniebla.com.mx/
Mexican permaculture training center.

North American Fruit Explorers
www.nafex.org
Cold-climate fruit geeks.
Northern Nut Growers Association
www.nutgrowing.org
Nut production enthusiasts.
Nuestras Raíces
www.nuestras-raices.org
Urban farming organization.
Perennial Solutions
www.perennialsolutions.org
Courses, workshops, and consulting
on edible landscapes and more.

Permaculture Activist
www.permacultureactivist.net
The North American permaculture
magazine and online resource.
Plants for a Future
www.pfaf.org
Phenomenal database of
useful plants.
USDA Plants Database
http://plants.usda.gov/java/
Detailed database of plants growing
in the United States.

SEED COMPANIES, NURSERIES, AND SUPPLIERS

Agroforestry Research Trust
www.agroforestry.co.uk
Forest garden species.
Baker Creek Seeds
www.rareseeds.com
Heirloom crops.
B&T World Seeds
www.b-and-t-world-seeds.com
Seed source for rarities from
around the world.
ECHO
www.echobooks.org
Tropical crops.
Edible Landscaping
www.eat-it.com
Fruits and nuts.
Evergreen Y. H. Enterprises
www.evergreenseeds.com
Asian crop seed.
Fedco
www.fedcoseeds.com
Vegetables, fruits, and more.
Food Forest Farm
www.permaculturenursery.com
Permaculture plant sales and
educational services.

Fungi Perfecti
www.fungi.com
Spawn of edible and useful fungi.
High Mowing Seeds
www.highmowingseeds.com
Organic vegetable seed.
Johnny's Selected Seeds
www.johnnyseeds.com
Wide range of annual crop seed.
Kitazawa Seed Company
www.kitazawaseed.com
Asian crop seed.
Logee's Greenhouse
www.logees.com
Tropical fruits for container or
greenhouse.
Meadow Creature Broadfork
www.meadowcreature.com
Compaction-busting tools.
Native Seed Search
www.nativeseeds.org
Native American crop seed varieties.
Nolin River Nut Nursery
www.nolinnursery.com
Grafted nuts, pawpaws, and
persimmons.

Oikos Tree Crops
 www.oikostreecrops.com
 Wide selection of useful trees
 and perennials.
One Green World
 www.onegreenworld.com
 Fruits and nuts.
Raintree Nursery
 www.raintreenursery.com
 Fruits and nuts.

Richters Herbs
 www.richters.com
 Useful medicinal and food plants.
Tripple Brook Farm
 www.tripplebrookfarm.org
 Useful plants, bamboos, and natives
 for permaculture and beyond.
USDA Germplasm Collection
 http://www.ars-grin.gov/npgs/
 Seed source of last resort for
 unusual species.

INDEX

Note: page numbers followed with c refer to photographs in the color insert

ABOUT THE AUTHORS

Eric Toensmeier has studied and practiced permaculture since 1990. He is the author of *Perennial Vegetables* and coauthor of *Edible Forest Gardens* with Dave Jacke. Toensmeier has worked as a small-farm trainer at the New England Small Farm Institute, has managed the Tierra de Oportunidades new-farmer program of Nuestras Raices, and is a graduate and former faculty member of the Institute for Social Ecology in Plainfield, Vermont. His current interest is in large-scale permaculture farming as a carbon-sequestering solution to climate change. Toensmeier's writing, consulting, and teaching business is based at www.perennialsolutions.org, where he posts his latest articles and videos. He lives in Holyoke, Massachusetts.

Jonathan Bates owns Food Forest Farm Permaculture Nursery (permaculturenursery.com), a nursery specializing in educational services and useful/edible plant sales. He's been studying, creating, and working with rural and urban gardens in the Connecticut River Valley for over a decade. With a bachelor's degree in biology and an MA in social ecology from the Institute for Social Ecology, Jonathan loves wildcrafting with friends and working with folks to better the world we live in. He cofounded and is a board member of the Apios Institute, is a teacher at the Yestermorrow Design/ Build School, and is a farmer with Nuestras Raices, Inc. He lives in Holyoke, Massachusetts.

"This logo identifies paper that meets the standards of the Forest Stewardship Council. FSC is widely regarded as the best practice in forest management, ensuring the highest protections for forests and indigenous peoples."

Chelsea Green Publishing is committed to preserving ancient forests and natural resources. We elected to print this title on 30-percent postconsumer recycled paper, processed chlorine-free. As a result, for this printing, we have saved:

16 Trees (40' tall and 6-8" diameter)
7 Million BTUs of Total Energy
1,373 Pounds of Greenhouse Gases
7,446 Gallons of Wastewater
498 Pounds of Solid Waste

Chelsea Green Publishing made this paper choice because we and our printer, Thomson-Shore, Inc., are members of the Green Press Initiative, a nonprofit program dedicated to supporting authors, publishers, and suppliers in their efforts to reduce their use of fiber obtained from endangered forests. For more information, visit: www.greenpressinitiative.org.

Environmental impact estimates were made using the Environmental Defense Paper Calculator. For more information visit: www.papercalculator.org.

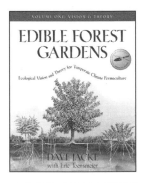
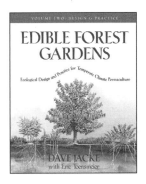

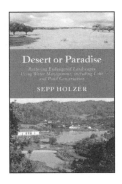

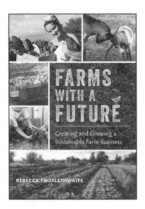
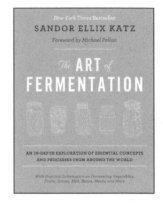